Geography, Ideology and Social Concern

edited by

D. R. Stoddart

Barnes & Noble Books
Totowa, New Jersey

©Basil Blackwell Publisher Ltd 1981
First published in the USA 1981 by
Barnes & Noble Books
81 Adams Drive
Totowa
New Jersey 07512

British Library Cataloguing in Publication Data

Geography, ideology and social concern.
  1. Geographical research - Political
  aspects - Congresses
  I. Stoddart, David Ross
  910'.7'2   G72.5

  ISBN 0–389–20207–X

Printed in Great Britain

Contents

Contributors   v

1   Ideas and Interpretation in the History of Geography   1
   D. R. Stoddart

2   The Contextual Approach   8
   Vincent Berdoulay

3   External Influence and Internal Change in the Development   17
   of Geography
   Olavi Granö

4   Institutionalization of Geography and Strategies of Change   37
   Horacio Capel

5   The Paradigm Concept and the History of Geography   70
   D. R. Stoddart

6   On People, Paradigms, and 'Progress' in Geography   81
   Anne Buttimer

7   Wilhelm Dilthey's Philosophy of Historical Understanding: a   99
   Neglected Heritage of Contemporary Humanistic Geography
   Courtice Rose

8   Peter Kropotkin, the Anarchist Geographer   134
   Myrna Margulies Breitbart

9   Elisée Reclus, an Anarchist in Geography   154
   G. S. Dunbar

10   Alfred Weber and Location Theory   165
   Derek Gregory

11   Geography and Social Science: the Role of Patrick Geddes   186
   B. T. Robson

12   Royce's 'Provincialism': a Metaphysician's Social   208
   Geography
   J. Nicholas Entrikin

13   Epistemology and the History of Geographical Thought   227
   Paul Claval

Index   240

Anne Buttimer's paper previously appeared in *Rapporter och Notiser* of the Institutionen för Kulturgeografi och Ekonomisk Geografi vid Lunds Universitet, 47 (1978), 1–25.

Paul Claval's paper first appeared under the title 'Epistemology and the history of geographical thought' in *Progress in Human Geography*, 4 (1980), 371–384.

Contributors

Vincent Berdoulay
Faculty of Arts, University of Ottawa
Dr Berdoulay earned his doctorate at the University of California, Berkeley, with a thesis on *The emergence of the French school of geography*, and he has published on the history of geography at the end of the nineteenth and beginning of the twentieth centuries.

Myrna Margulies Breitbart
School of Social Sciences, Hampshire College, Amherst, Massachusetts
Dr Breitbart's interest in Kropotkin stems from her research on Spanish anarchist movements in the 1930s.

Anne Buttimer
School of Geography, Clark University, Worcester, Massachusetts
In addition to her historical study of *Society and milieu in the French geographic tradition* (1971), Dr Buttimer has written widely on humanist approaches in geography, notably in her *Values in geography* (1974).

Horacio Capel
Department of Geography, University of Barcelona
Dr Capel has been largely responsible for the introduction of new approaches in Spanish geography, especially through the serial *Geo Critica* and the series of translations of English and American works which he has edited.

Paul Claval
Department of Geography, University of Paris
Professor Claval's prolific writings on the history and nature of geography include his *Essai sur l'évolution de la géographie humaine* (1964), *La pensée géographique* (1972), and *Les mythes fondateurs des sciences sociales* (1980).

G. S. Dunbar
Department of Geography, University of California, Los Angeles
Professor Dunbar is best known for his biography *Elisée Reclus: historian of nature* (1978); he has also written widely on the formative years of academic geography.

J. Nicholas Entrikin
Department of Geography, University of California, Los Angeles
Dr Entrikin's doctoral dissertation at the University of Madison, Wisconsin, was entitled *Science and humanism in geography*. He has published on the geographical implications of the work of Ernst Cassirer and Robert Park.

Olavi Granö
Department of Geography, University of Turku
Professor Granö has published in the history and epistemology of geography, mainly in Finnish and Swedish journals.

Derek Gregory
Sidney Sussex College, Cambridge
Dr Gregory is best known for his *Ideology, science and human geography* (1978). In addition to his historical and ideological interests he also works on nineteenth and early twentieth century historical geography.

B. T. Robson
School of Geography, University of Manchester
Professor Robson's main contributions include *Urban analysis* (1969), *Urban growth* (1973) and *Urban social areas* (1975). Before going to Manchester he was at Cambridge University.

Courtice Rose
Dr Rose lectures in the Department of Geography, Bishop's University, Lennoxville, Quebec.

D. R. Stoddart
Department of Geography, Cambridge University
In addition to his main work on the geomorphology and biogeography of coral reefs, Dr Stoddart is interested in the work of Darwin and Huxley and in the development of British academic geography at the end of the nineteenth century.

Ideas and Interpretation in the History of Geography

D. R. Stoddart

INTRODUCTION

This book has two purposes. The first is to demonstrate that the history of geography is more than simply the chronological listing of the achievements of a few great scholars arrayed in national schools, and that both the ideas and the structure of the subject have developed in response to complex social, economic, ideological and intellectual stimuli. The second is to show that throughout its recent history geographers have been not only concerned with narrowly academic issues, but have also been deeply involved with matters of social concern. The sharpening of the historical method and the redirection of its focus makes possible a new understanding of the activities and endeavours of those who have called themselves geographers.

The book had its origins in a joint meeting of the International Geographical Union's Commission on the History of Geographical Thought and of the International Union of the History and Philosophy of Science, in Edinburgh during August 1977, on the occasion of the International Congress of the History of Science. A colloquium under the title of 'The History of Geography and the History of Science' was organized around the themes of processes of change in geography, themes in its development, and the evolution of national schools. The present volume includes several of the papers given under the first of these two headings, together with additional contributions invited when the argument of the book had been developed during the meeting and after it.

In this introductory paper I wish to place both its methods and its themes within a wider framework of intellectual history, and to show how these essays differ in approach from the standard histories of our subject.

THE HISTORY OF GEOGRAPHY

In its simplest form, history is but a chronology of events. The main organizing device is narrative of change over time, and the archetype of such history in the field of geography is that of exploration. Baker (1931), to take but one example, provides a narrative listing over time of the progressive discovery of the lands and seas of the world. The story told is necessarily one of cumulative advance in knowledge towards the present, and in so far as there is any coherence to the story it is provided by the sequence of the events themselves.

It is a commonplace that under the influences of Kantian ideas, this methodology was readily transferred to the history of geography as a whole. Classic examples of works similarly organized are given by the recent books of Dickinson (1969, 1976). Such histories share common characteristics (Agassi, 1963). First, there is an emphasis on chronology, cumulation, and continuity. Scientific advance is seen as pressing relentlessly forwards towards the present, and issues such as priority in discovery or publication loom large as criteria of significance. Second, such inductivist history is markedly internalist. Written from the perspective of the present, such studies identify a continuous series of men and ideas, linked together in chronology and content. Thus Hartshorne (1939) and subsequently many others have traced a development within geography from Kant through Humboldt and Ritter to Richthofen and Hettner. Set squarely within what Schaefer (1953) termed the 'exceptionalist' tradition, it is hence not surprising that the names of Darwin, Marx and Freud are absent from Hartshorne's and similar works; that they give little or no attention to philosophical or epistemological issues; and that the history traced remains unrelated to social, economic and political conditions. Third, the identification of such a mainstream of progress involves value judgements about the past from the standpoint of what is clearly an evanescent present. Indeed Hartshorne (1939, chapter 3) devotes a section of his book to 'deviations from the course of historical development'.

Hartshorne, of course, was writing a methodological tract rather than making an historical enquiry alone, but his methods had profound consequences for historical writing in the subject and for the general perception of what was significant in our history. For him, history is normative: 'If we wish to keep on the track – or return to the proper track . . . – we must first look back of us to see in what direction that track has led' (Hartshorne, 1939, 31). Furthermore, the actors in the history are readily characterized into those who followed the track (and who were therefore right) and those who blundered off (and were hence wrong). The former heroes include men such as Varenius, Vidal de la Blache, Humboldt and Ritter. But such reputations are particularly vulnerable to changes in one's vantage point. Depending on the track one discerns, for example, one can see Carl Sauer either as hero or as villain: indeed the ambiguity in Sauer's position and the ambivalence with which he is often regarded stems largely from the difficulty with which he can be fitted into the standard unilinear pattern.

Finally, such narrative history says nothing of the processes of change. Ideas, in so far as they are discussed at all, are linked only by the simplest mechanisms of cause and effect, largely inferred from temporal succession. There is little consideration of individuals and their intellectual milieux; and such reconstructions hardly reflect the contingencies and pressures which all scholars recognize in their own intellectual development.

But there is another view of the history of the subject, recognized more than 50 years ago by John K. Wright (1926, 11), when he wrote that 'the history of geography . . . is the history of geographical ideas'. Sadly, Wright's call was scarcely heeded. Such a history will emphasize the development of problems and theories and the social and intellectual context of their protagonists, rather than the cataloguing of people, institutions and publications. The analysis of the history of ideas is necessarily contextual as Berdoulay argues in chapter 2. The truth or falsity of any particular proposition is no longer the prime criterion of its importance, whether from an absolute or a relative standpoint: indeed, questions must arise about the status of such categories themselves. Rather, we seek to understand how geographers as individual scholars recognized and grappled with intellectual issues in their time, in particular intellectual, social and economic environments. Can the history of geography in Europe between 1870 and 1914 begin to be understood without reference to the Franco-Prussian War and World War I? – to the educational expansion taking place in schools and universities? – to the Darwinian ferment, to which the French alone remained largely immune? – or to the emergence of the social and the human sciences in the eighties and nineties? Such a history simply cannot be written from the pages of the leading geographical journals of the time.

It follows that individuals themselves can no longer be categorized as 'good' or 'bad', and neither can their views. Wright (1926, 17) in the same prescient paper realized this clearly: 'Is not the history of error, folly, and emotion often as enlightening as the history of wisdom?' Agassi similarly in 1963 drew attention to the importance for scientific advance of 'clever mistakes' and 'stimulating error', but in geography only James (1967) has begun to explore this field.

The emphasis thus is on context and contingency as they affect the development of geographical thought. In the history of geography there are surprisingly few analyses of particular concepts. There are important papers by Plewe (1932) on comparative geography, by Bartels (1969) on 'harmony', and by Hard (1970) on *landschaft*. Others have explored the ways in which geographical ideas are embedded in and respond to a wider intellectual context. Lehmann in 1937 provided a remarkable but wholly overlooked consideration of the implications for geographical reasoning of the new statistical mechanics and atomic physics of Ernst Mach and Max Planck (though later commentators have perhaps wrongly claimed a greater influence for Heisenberg). There is too a growing literature on the impact of Darwin's ideas on the organization of knowledge and the methodology of explanation (Stoddart, 1966, 1975a, 1979). Lukermann (1965) in a brief but allusive paper has linked ideas in French geography at the turn of the century with the probabilistic work of Henri Poincaré and other mathematicians. But in general this is a field which remains largely unexplored.

PURPOSE

Hartshorne's justification for an interest in the history of geography was essentially as a guide to what one ought to do. The justification for contextual history is somewhat different. Can we recognize homologies in problems and responses between past and present? Can analysis of past dilemmas help us to cope with those which now beset us? But more, can we so deepen our understanding of the past that we cease to view its practitioners as two-dimensional figures, and recognize throughout the development of our subject the complexity and subtlety of the intellectual endeavour? Inductivist and internalist history bears a heavy responsibility for reducing these endeavours to such simplistic terms in our standard histories that their intellectual content has been demeaned, and our leading practitioners, while identified as the great men of the past, have been paradoxically automatically reduced in stature. And it is precisely in revealing the complexity of the past that we can throw light on the processes of historical change.

We could, of course, go further, and argue that these historiographical problems are basically epistemological: that they ultimately depend on what we understand knowledge to be. There is some discussion of these issues by Lowenthal (1961) and Kirk (1963), but only Gregory (1978) has made any sustained attempt to incorporate such issues into the fabric of the subject. There is undoubtedly an archaeology of geography, as Foucault has argued of knowledge in general, but it would be premature to attempt it here. Nevertheless, Claval in the concluding chapter, and other contributors, make some progress towards this aim.

GEOGRAPHY, IDEOLOGY, AND SOCIAL CONCERN

It is against this background that these essays are presented, as a contribution to a revisionist history of geography, and to further understanding of what geographers do and how and why they do it. This is not to dismiss or denigrate the standard histories of the subject, but it is to assert that we are dealing with an achievement more complex, more substantial and more rewarding than is immediately apparent from the narratives of James (1972), Dickinson (1969, 1976) and Freeman (1961).

The book deals with two substantive themes. First, we show the continuing commitment over many decades of geographers to social problems. The names of Kropotkin and Reclus, whose lives were disrupted by their principles, are perhaps best known, and are treated here by Breitbart and Dunbar. But we should not forget that in their concern for the human condition they mirrored the passionate reaction of Humboldt (and later Darwin) to the shame of slavery in Brazil. Robson shows, too, how geographers contributed to social planning half a century ago, at a time of widespread liberal alarm at the conditions of urban life. Thus the commitment to social relevance in geography has a long and distinguished history, which can only be dissected by patient reconstruction of past intellectual and social endeavours. We err if we think that passion and commitment have been newly discovered by modern scholars, as anyone who has read Kropotkin's paper of 1885 on 'What geography ought to be' (written in prison) will know (Stoddart, 1975b). Perhaps Febvre spoke to historians more than to us in his polemics, but it is tempting to substitute geography for history in some of his writings, and to realize that geographers too were going beyond a mechanical areal differentiation or a crude environmental determinism at the time at which he wrote that 'We have no history of Love. We have no history of Death. We have no history of Pity nor of Cruelty. We have no history of Joy' (Febvre, 1941, 18). Even to consider such categories transforms our conception of what it is that geography can do.

Second, we wish to make inferences about the nature and process of intellectual change. Berdoulay, Granö and Capel do this by setting change in a wider social and intellectual context. Gregory, Rose and Entrikin illuminate the issues with reference to the work of Weber, Dilthey and Royce, and readily demonstrate the inutility of existing generalizations. Finally, Buttimer and Stoddart explore the relevance of explanatory models of scientific change, with somewhat divergent results.

This book is, therefore, an exploration. It hopes to shed light not only on geography and what it is about, but on its history and how it should be written. Perhaps the historian of geography is at present in much the same position as the woodsman with his solitary light in the night in the forests of mediaeval Europe, in the celebrated introduction to Huizinga's *Making of the Middle Ages*. In this area, at least, the woods are still to be cleared and the fens to be drained.

ACKNOWLEDGEMENTS

I acknowledge first of all the initiative of Walter Freeman, secretary, and Philippe Pinchemel, chairman, of the International Geographical Union's Commission on the History of Geographical Thought in proposing the colloquium in Edinburgh in 1977. At the last moment I was myself prevented from attending, and my colleagues Derek Gregory and Mark Billinge served as local organizers, for which I am most grateful. Many of the contributors have shown great patience as the book has progressed towards completion. Two of our leading historians of geography, Clarence Glacken and David Lowenthal, have done much over the years to change our conceptions, both in publication and in conversation, and I am glad to acknowledge the stimulus of their work.

References

Agassi, J. 1963: Towards an historiography of science. *History and Theory*, Beiheft 2, 1–117.

Baker, J. N. L. 1931: *A history of geographical discovery and exploration* (London: 2nd edition 1937).

Bartels, D. 1969: Der Harmoniebegriff in der Geographie. *Die Erde* 100, 124–37.

Dickinson, R. E. 1969: *The makers of modern geography* (London).

Dickinson, R. E. 1976: *Regional concept: the Anglo-American leaders* (London).

Febvre, L. 1941: La sensibilité et l'histoire: comment reconstituer la vie affective d'autrefois? *Annales d'Histoire Sociale* 3, 5–20.

Freeman, T. W. 1961: *A hundred years of geography* (London).

Gregory, D. 1978: *Ideology, science and human geography* (London).

Hard, G. 1970: Die 'Landschaft' der Sprache und die 'Landschaft' der Geographen. *Colloquium Geographicum* 11, 1–278.

Hartshorne, R. 1939: The nature of geography: a critical survey of current thought in the light of the past. *Annals of the Association of American Geographers* 29, 171–658.

James, P. E. 1967: On the origin and persistence of error in geography. *Annals of the Association of American Geographers* 57, 1–24.

James, P. E. 1972: *All possible worlds: a history of geographical ideas* (Indianapolis).

Kirk, W. 1963: Problems of geography. *Geography* 48, 357–71.

Kropotkin, P. 1885: What geography ought to be. *The Nineteenth Century* 18, 940–56.

Lehmann, O. 1937: *Der Zerfall der Kausalität und die Geographie* (Zurich).

Lowenthal, D. 1961: Geography, experience, and imagination: towards a geographical epistemology. *Annals of the Association of American Geographers* 51, 241–60.

Lukermann, F. 1965: The 'Calcul des probabilités' and the 'Ecole Française de Géographie'. *Canadian Geographer* 9, 128–37.

Plewe, E. 1932: Untersuchung über die Begriff der 'Vergleichenden' Erdkunde und seine Anwendung in der neueren Geographie. *Zeitschrift der Gesellschaft für Erdkunde zu Berlin*, Ergänzungsheft 4, 1–92.

Schaefer, F. K. 1953: Exceptionalism in geography: a methodological examination. *Annals of the Association of American Geographers* 43, 226–49.

Stoddart, D. R. 1966: Darwin's impact on geography. *Annals of the Association of American Geographers* 56, 683–98.

Stoddart, D. R. 1975a: 'That Victorian science': Huxley's *Physiography* and its impact on geography. *Transactions of the Institute of British Geographers* 66, 17–40.

Stoddart, D. R. 1975b: Kropotkin, Reclus, and 'relevant' geography. *Area* 8, 188–90.

Stoddart, D. R. 1979: *Darwin's influence on the development of geography in the United States, 1859-1914*. (Conference on the Evolution of Academic Geography in the United States, Lincoln, Nebraska, April 1979).

Wright, J. K. 1926: A plea for the history of geography. *Isis* 8, 477–91; reprinted in J. K. Wright, *Human nature in geography* (Cambridge, Mass., 1966), 11–23.

2

The Contextual Approach

V. Berdoulay

The history of geography is a repository of ideas about the relationship between man and nature. It is an account of man's experience in trying to comprehend his world. In other words, it greatly reflects the development of human consciousness. It is thus understandable that the history of geography can be ideologically loaded and given various and conflicting interpretations. All this points to a growing desire among geographers to better understand the actual context of the development of geographical ideas and institutions. However, the methodology of approaching such a concern has not been clarified, and the following pages are aimed at contributing to the elaboration of a contextual approach to the history of geography. The importance of looking at the context of geographic thought is first underlined. Then the difficulty of bringing the context to bear on the development of geography is examined in the light of the historiography of science so that, finally, the nature of a contextual approach is outlined in its main features.

One often reads that a geographer was 'influenced' by another. This question of so-called influence may be quite misleading. Actually, an approach based on this type of concern runs the danger of being plagued by all the pitfalls associated with the emphasis on the linear evolution of ideas. It rests on assumptions which philosophers and historians of science have easily criticized for their positivistic bias (Agassi, 1963; Hahn, 1965). Amazingly enough, these assumptions remain widespread among the public at large and even among scientists themselves.

These assumptions involve the belief in a continuous development of science by accumulation of facts, discoveries, and knowledge in science. The task, then, is to trace the progressive victory of truth over error, of 'good' ideas over 'bad' ideas, and of the inevitable emergence of true scientific ideas from facts. The historian of science is thus concerned with whom to worship. Error is viewed as something wicked which hinders the development of science; consequently, little interest is paid to historical contexts or intellectual climates since the focus is placed on the internal evolution of each science.

These naively positivistic stances are nevertheless present among geographers. For instance, the variety of geographic research trends in the second half of the nineteenth century is rarely regarded as a significant issue. In fact, the trend which became dominant in the university system – such as that of Vidal de la Blache in France – is considered the expression of the mainstream of the best research, as the 'good', the 'superior' trend (e.g. Meynier, 1969 vs. Claval and Nardy, 1968). The past is thus explained in the light of the present: this is a teleological interpretation which ignores or neglects the problem of explaining the relative failure of other trends. The works of Richard Hartshorne (1939, 1968) – however thoughtful they may be – provide a good example of the pitfalls of this approach. Once the author has identified what the 'good', or 'right', idea of geography is, he traces it back to Kant, Humboldt and Ritter. The other trends which do not conform to this model are simply viewed as deviations. V. A. Anuchin's work (1977) is marred by a similar approach. It shows the victorious march of the (right) materialist concept of geography over idealist (thus wrong) developments in the discipline.

This type of approach has remained widespread among practitioners of geography, as is often evidenced in the introductory paragraphs of articles by authors who were associated with the 'quantitative revolution'. It is stated that the construction of theories or models based on quantitative techniques is the only valid approach since it allows for a rapid accumulation of knowledge and since it rests on a rigorous, value-free, and logical inner development of the discipline. In all the above examples, the history of geography is interpreted and distorted in order to justify particular methodological or epistemological positions.

Other approaches have tried to be pluralistic: several geographic 'traditions' or 'streams' are identified and their respective evolution is traced throughout history (e.g. Pattison, 1964; Haggett, 1965). Some perspective, however, is provided by the fact that a similar approach has been used in the history of science and that its weaknesses have been demonstrated. Developed at the turn of the century in Europe, particularly in France, by authors such as Henri Poincaré (1902, 1905) and Pierre Duhem (1906), this approach is sometimes referred to as 'conventionalism' in the sense that it stressed the conventional nature of theories. It was tied to a revival of Kant's philosophy whereby criticisms of the positivistic presuppositions of the above-described approach were introduced. In this respect, it had the merit of rejecting the idea that some trends are scientifically superior to others, and the empiricist belief that scientific theories emerge solely out of facts. The latter, according to this philosophy, were classified into particular theories on the basis of simplicity and convenience. It followed that, for the historian of science, the emphasis was still on the continuous development of ideas, but all past scientific trends were considered worthy of investigation.

Although these views on the history of science were more sophisticated than the previous ones, they kept ignoring the issues of discontinuities in the evolution of ideas, the interplay among the various scientific trends or traditions, the actual conditions of what research was about in the past (there was no way to predict the future success of competing theories), and the factors of change which were not 'internal' to science. This is not to say that studies in the filiation of specific ideas (explicitly taken out of their context) may not be extremely rewarding. The works by Arthur Lovejoy (1942) or Clarence Glacken (1967) clearly demonstrate the contrary. They show the persistence of a few ideas in the history of the Western world and, in so doing, highlight them in a fresh perspective. Nevertheless, this approach is not designed to analyse scientific systems of thought of a relatively short period of time, whereas it is often necessary to do so in the history of geography.,

Actually, it is felt more and more that factors 'external' to the development of geographic thought should be given due consideration (Hooson, 1968; Claval, 1972; Granö, 1977). This is, in fact, in line with contemporary epistemological concerns (Thuillier, 1972). There remains, however, the need to define an adequate methodology. The lessons that one may draw from the historiography of science are quite useful in this respect. Some of its most relevant trends are now examined.

One approach has been to stress the role of the *Zeitgeist*, which is considered as determining the way scientists and intellectuals view and deal with the world – as J. Burckhardt (1860) did in his famous book on the Italian Renaissance. This approach has the merit of showing the interdependence of a science with the thought of the same epoch. But it has been criticized for its tautological bias and thus for its weak explanatory power. Moreover, this approach is too remote from the internal logic of science, the interplay of small-scale theories, hypotheses, and techniques. The world view of a scientist provides an explanation for his frame of reference but cannot justify all the specific initiatives taken by him. This approach thus cannot account for the numerous research trends which survive, or rise, at odds with those which correspond to the *Zeitgeist*. This may be why, in the history of geography, the resort to the *Zeitgeist* has rarely gone beyond the level of brief hints: it would not easily resist a thorough examination of the variety of research trends at any epoch. Nevertheless, research on such linkages remains desirable and, in this respect, a few remarkable attempts have been made (e.g. de

Dainville, 1940; Lukermann, 1965; Büttner, 1973).

Another approach, which bears more similarities to the previous one than one would think at first, consists in stressing the socio-economic conditions of an epoch in order to explain the emergence of new scientific ideas. Numerous contributions, usually of Marxist inspiration, have been made since the inter-War period (e.g. Bukharin *et al.*, 1931; Haldane, 1939; Bernal, 1954). This approach has the merit of asking new questions, of exploring the place of science within the total structure of society. But a similar criticism to the *Zeitgeist* approach can be made, that is, the difficulty of establishing precise relationships between general (socio-economic) conditions and specific, internal developments of a science. The numerous and strong attacks against Hessen's famous attempt (1931) to explain Newtonian science by its socio-economic conditioning is a good example in point (see Butterfield, 1950).

Here again histories of geography have evidenced a great deal of caution in these matters, although there is no lack of hints, especially with respect to the impact of colonialism or capitalism on geographic thought. The essays by Tricart (1953, 1956) and Lacoste (1976) are among the rare examples of the application – albeit incomplete – of this type of approach and actually suffer greatly from the sweeping generalizations which are found in them. In order to avoid the simplistic determinism which tends to plague these works, a solution is to limit the approach to separating ideas (or the inner logic of science) from socio-economic conditions. The latter are then viewed as mere barriers to the 'normal' development of science (e.g. Needham, 1954-). But this boils down to making a strict distinction between 'external' and 'internal' factors in scientific evolution, that is, to base the dynamics of science on shaky and much decried assumptions (Lécuyer, 1968; Kuhn, 1968). In addition, this constraining effect of external factors on internal ones is reminiscent of the 'stop-and-go determinism' which was defended without much success by Griffith Taylor (1957) and thus does not incite the historian of geographic thought to turn toward this type of approach.

Recent contributions to the interrelatedness of ideas in both science and society have been made by setting forth the human dimension of science. This has been achieved by emphasizing the existence of discontinuities in scientific evolution (Bachelard, 1953; Koyré, 1939), or the social and cognitive nature of scientific research (Piaget, 1967; Kuhn, 1970; Gusdorf, 1966; Foucault, 1966, 1969). Thomas Kuhn's interpretation and presentation of these ideas have been popular among geographers (e.g. Chorley and Haggett, 1967; Berry, 1973) although their comprehensive application to the history of geography is lacking. This should not come as a surprise as there are too many epistemological and historical difficulties with the very notion of paradigm (Lakatos and Musgrave, 1970) to hope for an application of Kuhn's ideas to a specific, limited field of investigation (e.g. Edge and Mulkay, 1976), especially in the non-natural sciences (which were not considered in Kuhn's famous book). Nevertheless, the pertinence of Kuhn's contribution was to enhance the social-cognitive nature of science and to propose a model of its social dynamics. It stops short, however, of providing a lead to place science in its total societal context. Other contributions come closer to this objective, such as those by Gusdorf on the 'models of intelligibility' of an epoch or by Foucault on the 'épistémé' which underlies scientific activity. But this is at the expense of what is, in my opinion, Kuhn's most important contribution, that is, his concern for the social dynamics of scientific change.

In order to encompass the above-mentioned perspectives, it is necessary to take into account all the sociological determinations of scientific knowledge. Along these lines, some assistance is offered by the sociology of science. It is a relatively recent field of study which, through the study of scientific communities, resistance to innovations, and the organization of scientific education and research in various countries, attempts to understand what are the most favourable conditions for the development of science (Merton, 1973; Ben-David, 1970). The convergence of these interests with those of policy makers and historians of science have led to a host of new publications on what is often termed the science of science or the social study of science (Hahn, 1975). It thereby raises interesting questions which have been hardly touched on in the literature on the history of geography, such as the uneven status of geography among countries, the state of geography in 'developing countries', the diffusion of geographic concepts across boundaries, the shift of innovative centres from one country to another through time, and the role of various institutions in influencing the orientation of geographic research. However, there is a deliberate emphasis on purely external conditions of scientific development and the same criticisms can be levelled at this approach as were made about the Marxist-inspired works discussed above. A substantial number of scholars are now trying to get away from this 'externalist' bias (Lécuyer, 1976; Hahn, 1975). More attention is paid to the content of scientific knowledge and its dependency on social processes is investigated. In other words, it is scientific knowledge itself, and not only its rate or direction of growth, which is viewed as a social product (Mulkay, 1978).

One aspect of the sociology of science – the study of the institutionalization of innovations – may be treated as a valuable intermediary link in the relations between the development of geographic thought and its societal context (Berdoulay, 1980). Relevant research – albeit rather 'externalist' – on the social sciences has been done by authors such as Ben-David (1970, 1971), Oberschall (1965), and Clark (1968, 1973). Attention is thus drawn to such considerations as the factors influencing the degree of acceptance of an innovation, the establishment of professional activities (e.g.

journals), the definition of the new status associated with the innovation, the stages of institutionalization, and the presence of competing groups of scientists. The study of institutionalization – as a process – helps to bring the research focus on the conjunction of internal with external factors.

In light of the above review and discussion of some relevant historiography, a contextual approach can be formulated along a set of methodological guidelines as follows.

Firstly, there are two fundamental assumptions on which this approach rests. The first one is that there exist changing systems of thought at the same time as there is continuity of certain ideas. But this view does not imply that a system of thought should be *a priori* assigned to each historical period or to each social group. The second assumption is no more demanding: there is no radical dichotomy between internal and external factors of scientific change. These factors may be viewed only as two points on a continuum, without any sharp distinction. It must be kept in mind that these two assumptions have received considerable historical as well as philosophical support (e.g., see Toulmin, 1972).

Secondly, the contextual approach should not neglect any geographic trend, even if some of them acquired no posterity. The point is to start research without assigning any intellectual superiority to some trend or another. One may indeed discover later that the reasons behind a lack of success or posterity may be essentially sociological or political. It is certainly in this respect that most histories of geography are biased. They take a quasi-'official' point of view which reinforces the preconceptions or prejudices of the past. Thus, not only are interesting works ignored but also the historian of geography deprives himself of elements which would permit him to understand the basic significance of the trend which gained primacy. It is precisely this type of contextual concern, for instance, which allowed me to uncover the epistemological depth of Vidal de la Blache's contribution to geography (Berdoulay, 1976, 1978).

Thirdly, the identification and in-depth study of the major issues which concern a society are necessary, even if some of them may not seem, at first glance, to have influenced the evolution of geographic ideas. For example, the much neglected issue of ethics in a changing society has been shown to be of primary importance in the shaping of French geography at the turn of the century. Only the type of reflections which this issue promoted and the type of answer which was proposed permit the understanding of the social geographical views of the concerned French scholars (Berdoulay, 1974). Fresh areas of research are thus opened by a long-unrecognized mainspring of geographic thought.

Fourthly, as geographic trends have some sociological basis, it is important not to adopt a concept of 'scientific community' as narrow as is often found in the sociology of science. The identification of linkages among scientists alone is not sufficient in explaining the context of geographic research or the existence of diverse trends. It is imperative to put more emphasis on ideologies than on institutions proper. Any scholar belongs to what can be termed a 'circle of affinity' which encompasses more than a scientific community. It includes specialists of very diverse disciplines as well as politicians or intellectuals whose positions on societal issues are known. This is the only way to cast light on the ideas of geographers who seem isolated but whose circle of affinity is most revealing. What is significant in order to understand their geographic thought is not so much their lack of contact with a community of geographers as the ideological leanings which brought them into contact with non-geographers. Striking examples are those of Elisée Reclus whose circle of affinity included, among others, numerous anarchists, and of Emile Levasseur whose links with groups of economists, statisticians, and social reformers have been misappreciated (see Nardy's contribution in Claval and Nardy, 1968, and Berdoulay, 1974).

Finally, the approach consists less in examining the possible 'influence' of an idea than in looking at the reasons behind the 'demand' for, or 'use' of, this idea. A similar conclusion is arrived at by Toulmin (1972) in his philosophical reflections on knowledge. The context then best explains the originality of the synthesis of a particular set of ideas held by an individual or a group, while any one of these ideas, taken separately, may not be new or innovative.

The contextual approach, hardly formalized as it is, serves as a comprehensive framework for analysing the conjunction of the inner logic and content of science and the context in which the scientist is placed. It is by disentangling the links which unite change in geographic thought to its context that one is in the best position to assess, and to learn from, the creative contributions of great individuals.

References

Agassi, J. 1963: Towards an historiography of science. *History and Theory*, Beiheft 2, 1–117.
Anuchin, V. A. 1977: *Theoretical problems of geography* (Columbus, Ohio), translation of *Teoreticheskie problemy geografii*, 1960.
Bachelard, G. 1953: Le Matérialisme rationnel (Paris).
Ben-David, J. 1970: Introduction. *International Social Science Journal* 22(1), 7–27 (special issue on the sociology of science).
Ben-David, J. 1971: *The scientist's role in society. A comparative study* (Englewood Cliffs).
Berdoulay, V. 1974: *The emergence of the French school of geography*. Ph.D. dissertation, University of California (Berkeley).
Berdoulay, V. 1976: French possibilism as a form of neo-Kantian philosophy. *Proceedings of the Association of American Geographers* 8, 176–79.

Berdoulay, V. 1978: The Vidal-Durkheim debate. In Ley, D. and Samuels, M., editors, *Humanistic geography. Prospects and problems* (Chicago) 77–90.

Berdoulay, V. 1980: Professionnalisation et institutionnalisation de la géographie. *Organon* 14.

Bernal, J. D. 1954: *Science in history* (London).

Berry, B. J. L. 1973: A paradigm for modern geography. In Chorley, R. J., editor, *Directions in geography* (London) 3–21.

Bukharin, N. I. *et al.* 1931: *Science at the cross roads* (London)

Burckhardt, J. 1860: *Die Cultur der Renaissance in Italien, ein Versuch* (Basel).

Butterfield, H. 1950: *The origins of modern science* (New York).

Büttner, M. 1973: *Die Geographia Generalis vor Varenius:geographisches Weltbild und Providentiallehre.* (Erdwissenschaftliche Forschung 7, Wiesbaden).

Chorley, R. and Haggett, P. 1967: Models, paradigms and the new geography. In Chorley, R. and Haggett, P., editors, *Models in geography* (London) 19–41.

Clark, T. N. 1968: Institutionalization of innovation in higher education: Four models. *Administrative Science Quarterly* 13, 1–25.

Clark, T. N. 1973: *Prophets and patrons: the French university and the emergence of the social sciences* (Cambridge, Mass.).

Claval, P. 1972: *La pensée géographique. Introduction à son histoire*(Paris).

Claval, P. and Nardy, J.-P. 1968: *Pour le cinquantenaire de la mort de Paul Vidal de la Blache* (Paris).

Dainville, F. de 1940: *La géographie des humanistes* (Paris).

Duhem, P. 1906: *La théorie physique, son objet et sa structure* (Paris).

Edge, D. O. and Mulkay, M. J. 1976: *Astronomy transformed* (New York).

Foucault, M. 1966: *Les mots et les choses* (Paris).

Foucault, M. 1969: *L'archéologie du savoir* (Paris).

Glacken, C. 1967: *Traces on the Rhodian shore. Nature and culture in Western thought from ancient times to the end of the eighteenth century* (Berkeley and Los Angeles).

Granö, O. 1977: Geography and the problem of the development of science. *Terra* 89 (1), 1–9 (Finnish and English texts).

Gusdorf, G. 1966: *De l'histoire des sciences à l'histoire de la pensée* (Paris).

Haggett, P. 1965: *Locational analysis in human geography* (New York).

Hahn, R. 1965: Reflections on the history of science. *Journal of the History of Philosophy* 3, 235–42.

Hahn, R. 1975: New directions in the social history of science. *Physis* 17 (3–4), 205–18.

Haldane, J. B. S. 1939: *The Marxist philosophy and the sciences* (New York).

Hartshorne, R. 1939: The nature of geography: a critical survey of current thought in the light of the past. *Annals of the Association of American Geographers* 29, 171–658.

Hartshorne, R. 1968: The concept of geography as a science of space from Kant and Humboldt to Hettner. *Annals of the Association of American Geographers* 48, 97–108.

Hessen, B. M. 1931: The social and economic roots of Newton's *Principia*. In Bukharin, N. I. *et al.*, *Science at the cross roads* (London, 1931) 151–76.

Hooson, D. 1968: The development of geography in pre-Soviet Russia. *Annals of the Association of American Geographers* 58, 250–72.

Koyré, A. 1939: *Etudes galiléennes*, 3 vols. (Paris).

Kuhn, T. 1968: History of science. *International Encyclopedia of the Social Sciences* (New York) vol. 14, 74–83.

Kuhn, T. 1970: *The structure of scientific revolutions* (Chicago: second edition, with a postscript).

Lacoste, Y. 1976: *La géographie, ça sert, d'abord, à faire la guerre* (Paris).

Lakatos, I. and Musgrave, A., editors, 1970: *Criticism and the growth of knowledge* (Cambridge).

Lécuyer, B.-P. 1968: Histoire et sociologie de la recherche sociale empirique: problèmes de théorie et de méthode. *Epistémologie sociologique* 6, 119–31.

Lécuyer, B. -P. 1976: Le système mertonien de sociologie de la science: genèse, ramifications, contestation. *Bulletin de la Société Française de Sociologie* 3 (6), 3–21.

Lovejoy, A. 1942: *The great chain of being: a study in the history of an idea* (Cambridge, Mass.).

Lukermann, F. 1965: The 'Calcul des probabilités' and the 'Ecole Française de Géographie'. *Canadian Geographer* 9, 128–37.

Merton, R. K. 1973: *The sociology of science*, edited and with an introduction by N. W. Storer (Chicago).

Meynier, A. 1969: *Histoire de la pensée géographique en France (1872–1969)* (Paris).

Mulkay, M. J. 1978: Consensus in science. *Social Science Information* 17(1), 107–22.

Needham, J. 1954– : *Science and civilisation in China*, several vols (Cambridge).

Oberschall, A. 1965: *Empirical social research in Germany, 1848–1914* (Paris and The Hague).

Pattison, W. D. 1964: The four traditions of geography. *Journal of Geography* 5, 211–216.

Piaget, J., editor, 1967: *Logique et connaissance scientifique* (Paris).

Poincaré, H. 1902: *La science et l'hypothèse* (Paris).

Poincaré, H. 1905: *La valeur de la science* (Paris).

Taylor, G., editor, 1957: *Geography in the twentieth century* (London).

Thuillier, P. 1972: *Jeux et enjeux de la science. Essai d'épistémologie critique* (Paris).

Toulmin, S. 1972: *Human understanding*, vol. 1 (London).

Tricart, J. 1953: Premier essai sur la géomorphologie et la pensée marxiste. *La Pensée* 47, 62–72.

Tricart, J. 1956: La géomorphologie et la pensée marxiste. *La Pensée*, 69, 3–24.

3

External Influence and Internal Change in the Development of Geography

Olavi Granö

The portrayal of how geography has developed must always be a reconstruction. The study of the history of geography depends essentially on how what is today defined as geography is regarded at any one stage in its development in the light of contemporary criteria. The domain of geography as a body of knowledge varies greatly today and it has also fluctuated greatly in the course of history. Consequently, any historiography of geography will vary according to the way in which geography as a concept is seen. The situation becomes quite different if geography is regarded as an institutionalized academic discipline, in which case its history is clearly delineated. The selection and moulding of knowledge to the domain of a given social institution is a decisive event in the development of geography. This paper considers the background and general character of this process of change, namely geography's development into a formal academic discipline, without slavishly adhering to any one country's tradition. References to concrete events do not indicate regional differences in the development of geography, the 'geography' of geographical research. They are only empirical examples with the help of which an attempt is made to indicate the trends in the development of geography and the overall background.

1 RESEARCH PRAXIS AS A LINK BETWEEN SCIENTIFIC KNOWLEDGE AND SOCIAL STRUCTURE

The branch of research that has adopted science as its object of study has focused its attention to an increasing extent on science as a social institution. Science is no longer regarded simply as a reflection of the object of study but also as a part of society. As a result, there have been a great number of studies on the sociology and social history of science in the wake of the classic works in this field by Merton (1938) and Bernal (1939). In the seventies, this research has swelled to take on the proportions of a flood (see e.g. Mulkay, 1977; MacLeod, 1977). This is quite understandable, for science has come to constitute such an important part of society that merely as a social phenomenon it deserves to be studied. Seen from the point of view of the development of science, the question of whether the external structure of society has influenced the content of science is also centrally important. Has science developed along intellectual lines quite independently of society or has it been affected? If so, to what degree? Because of this dichotomy, a distinction has been made between internal and external factors: for a discussion of this see Gustavsson (1971), Spiegel-Rösing (1973), Andersson (1975) and MacLeod (1977).

The link between the content of science and its social structure is *research praxis*, which constitutes part of the action of scholars. By action is here meant both their cognitive action and their concrete instrumental activities. As a result of the view that science is essentially research there arose two schools of thought: first, a phenomenology and hermeneutic of science (Heidegger, Kockelmans, etc.); second, the Polish 'praxiological' approach, based on the theory of purposive-rational action (Kotarbiński, etc.) and associated with the first interdisciplinary studies of the research process (Ossowska & Ossowski). These schools developed unnoticed as early as the 1930s (Radnitzky, 1974, 730). It was only after the publication of Kuhn's work (1962) that more general consideration began to be given to the socio-psychological contexts of the scientific theories prevailing at each stage in the history of science (cf. e.g. Böhme, 1977). Research praxis came to be seen as the achievement of the scholar's intentional action and therefore connected with the social environment. This brought about the union of the internal cognitive history and the external social history of science. As a result there have now emerged new 'humanistic' tendencies in the history of science, which stress the human implications of scientific advance (see Spiegel-Rösing, 1977).

On this basis, therefore, geography can be viewed in three ways: (1) an originally unorganized body of knowledge from which has developed the *scientific content* of geography, (2) practical action, from which *research praxis* has evolved, and later (3) a social institution, a *discipline*. No distinction should be made between these three since knowledge and the increase in knowledge are bound to the individual scholar, who in turn belongs to a given social group. The variations that occur in these three components over time constitute the history of geography. Seen in this light, geography has become a larger totality so that as knowledge it belongs to science in the broad sense of the word ('Wissenschaft'), as research praxis it constitutes part of general scientific practice, and as a social institution it forms part of the social and organizational structure of science. Science in its entirety, however, constitutes as knowledge part of culture in general, as research praxis part of human action and behaviour, and as a social institution part of the social structure of society (figure 3.1).

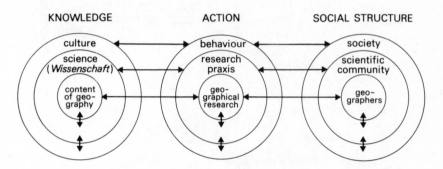

FIGURE 3.1   The context of geography

2 THE FOUNDATIONS OF HISTORIOGRAPHY

Figure 3.2 shows one way in which the historiography of geography might be represented diagrammatically. Different stages of time in geography, an earlier stage (a) and the present time (b), are placed on Hägerstrand's time axis (see e.g. Lenntorp, 1976) representing the succession of events in time. On the left hand side is the history of knowledge in geography as it has evolved from scientific ideas and theory; the aim of this, in accordance with science policy, is the creation of a geographical 'Weltbild'. On the right hand side is research praxis showing the action of different periods of time; the aim here is to transform knowledge. Surrounding this there is the social structure, the temporal changes of which show the social history of geography. The plane which unites the 'subject' and the 'real object' into a hermeneutic unit is an attempt to portray the linking of research praxis and geographical knowledge. According to this, man receives information from the real world as perception and experience, and he is on the other hand capable of action. The bridge between perception and action is formed by man's internal world of knowledge. The person is portrayed as constituting some kind of *transformation juncture* where, on the one hand, the perception of the flow of information is transformed into knowledge and action and, on the other, the future becomes the past. Consequently, at each transformation juncture (1) the acquisition of knowledge, (2) the application of knowledge, (3) the reconstruction of development (the historian's view) and (4) the

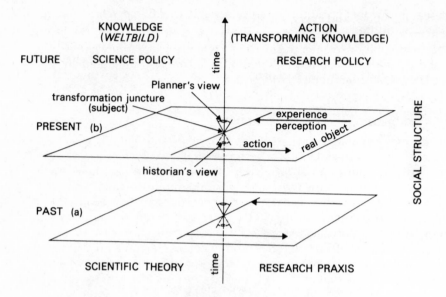

FIGURE 3.2   'Historiography' of geographical enquiry

programming of the future (the planner's view), are all the product of the same kind of praxis and often of the same person.

However, because of the time difference there is an essential difference between the praxis of the historian attempting a reconstruction of the past of geography and the praxis directed towards contemporary geographical research. The study of history may take place in quite a different place and under quite different conditions from those prevailing at the time which constitutes the period of study. The historian uses terms that have been created at a later period or he may define old terms in a new way. Historical development is divided into periods defined in a special way, and the people living during those periods had no idea that they were living in an intellectual or social framework that would later be described as a given period. The historian, as he looks back in time, reconstructs the praxis of the past time that constitutes his object of study at a later date when research praxis and its aims are quite different. As the 'now' level, the transformation juncture, moves upward along the time axis a continual series of transformations occurs. This change is brought about by all the research events of the intervening period, events which themselves have contributed to the development of geography during the period. The more significant the research event, the more it will have changed research praxis with the aid of which we later try to reconstruct the event.

Writing the history of geography is therefore closely linked to the research praxis of the time and place in which it is written. It might be said that every epoch and every school of research has its own historiography and compiles its history of geography in its own way (Hard, 1973a, 77). The future may also enter the picture in that historiographical surveys do not only portray past development but they are also programmatic and lay the foundations of new strategies.

3 THE ORIGIN OF THE HISTORIOGRAPHY OF GEOGRAPHY

Historiography may be reconstructive to the extent that only when history is written is the totality whose development we are trying to describe created and then in an artificial way. Put in another way the research praxis that evolves in the writing of a historiography did not exist earlier. This was the case when geography was becoming institutionalized as a discipline in Germany in the 1870s. The first writers of the history of geography therefore had to write about a discipline which had not existed up to then (Beck, 1954). Histories written before geography had attained the status of a discipline were simply descriptions of explorations (the best known being that of Peschel, 1865) or they dealt only *en passant* with the history of geographical thought (e.g. Lüdde, 1849). This shows that before geography became a discipline it was regarded as a branch of knowledge dealing with exploration journeys and research travels and that no link with the 'geographical' thinking of antiquity or the eighteenth century was explicitly mentioned.

However, once it had been established as a discipline, the history of geography was written in such a way that contemporary geographical research was projected into the past. The first university geographers tried to create geographical tradition. Writing the history of geography was part of the search for identity. Links with the geographical thinking of antiquity, comprising not only cartographic but also chorographical and topographical traditions, were gradually established as geography took on the status of a discipline. Among the most notable of the influences on geography at this time were von Richthofen's inaugural lecture in 1883 and Berger's detailed study of geography in antiquity (1887). Attention was first drawn to the development of geography and to its interrelations with surrounding society only in Wisotzki's work (1897). A return to this link was not then made until the 1960s.

When geography was still in its early stages as a discipline many surveys of the nature of geography were published, for it was frequently a favourite topic of inaugural lectures. In these discussions there were attempts to reconstruct the history of geography. The development of geography was often regarded as a continuous unilinear and cumulative process, a progression from achievement to achievement towards an independent science with its own object of study. The existence of a separate discipline called geography was thought to be self-evident, 'given', and an integral part of the development of geographical knowledge. The reconstruction of developments before the discipline stage varied from writer to writer. It was made up not only of the teaching tradition of the old universities and schools but also of such widely differing fields as the achievements of cartographers, surveyors, merchants, explorers and adventurers. The link with the history of science in general was that certain outstanding men of science, the list of whom varied according to the writer, were looked upon as geographers. However, when geography had become institutionalized as a discipline the situation changed completely. The history of geography was no longer a complete reconstruction in so far as it was now possible to recognize a geographer from among others. The content of these persons' research work was, at the same time, the history of geography.

It is, then, possible to identify two periods in the development of geography. The structure and development *per se* of these two periods is quite different: (1) the undefined pre-disciplinary period characterized merely by the existence of geographical knowledge, and (2) the stage at which geography had achieved the status of an institutionalized academic discipline. This latter stage was characterized by its own geographical research praxis that had been formed in one way or another, and could clearly be defined as a special field of knowledge. The history of the former period has been written principally as a reconstruction in relatively recent times.

4 THE STRUCTURE AND PROCESS OF GEOGRAPHICAL ENQUIRY

It is possible to establish old pre-disciplinary and pre-scientific traditions for geographical thought. Acknowledgement of the fact that geography is as old as human existence enables us to understand geography phenomenologically.

> Geographical science has in fact a phenomenological basis; that is to say, it derives from a geographical consciousness. On the one hand, the geographer develops this consciousness and makes society more aware of geography, but on the other hand the rise of geographical science is dependent on the existence of a pre-scientific and natural geographical consciousness . . . Geographers and geographical science can only exist in a society with a geographical sense. (van Paassen, 1957, 21)

This phenomenological view may be used to describe and evaluate the whole development of geography, even of later periods, although the studies themselves have been 'positivistic'. May (1970, 23–4) refers to this in pointing out the phenomenological background to geography during the years between the wars. On the other hand, a 'humanistic' movement has grown up in geography in recent years, and this movement has explicitly stated that its philosophical background is phenomenological or existential (Tuan, 1971, 1974; Buttimer, 1974, 1976; Relph 1970, 1976; Samuels, 1971).

Geographical knowledge, whatever that has meant, has been continually growing and changing. The enquiry for geographical knowledge, however, is based on a certain basic thinking independent of transformations in time and the fragmentation of the field of knowledge. This pre-scientific thinking has always existed as a covert parallel 'sub-scientific line' throughout the history of geography. The origin of geographical knowledge in the very earliest days of culture as expressed later by the oft-posed questions 'Where is it?' and 'What is it?' is based on man's consciousness of his environment. Man together with his environment constitutes an intertwined entity from which neither man nor the environment can be separated. Attempts to break this unity because of the subjective element involved, and to make man and his environment quite separate objects of logico-empirical study, have led to considerable problems for geography as a discipline. These problems in turn have been reflected in the often posed question of the unity of geography, monism, i.e. does geography as a discipline have an internal cognitive unity, or is it just a social structure of organizationally compatible branches of knowledge from neighbouring disciplines? This problem is seen most clearly in the dualism between physical and human geography.

The dilemma of the unity of geography is not, however, a question of the relation between man and nature as the *object* of study, a question of explaining man in terms of nature, nor of their integration as landscapes, regions, etc. Rather it is the question of the reciprocity between man's mind, the *subject*, and his environment, an attempt to explain land and nature in terms of man. Man's perception, experience, knowledge and action form together with his environment a totality, a unity, which constitutes the basic premise of geographical enquiry and which is illustrated in figure 3.3. Here experience is isolated from knowledge, and is formed through immanent sensation of the environment. Action is also extrapolated from knowledge and by different means brings about transformation of the environment. These, experience and action, correspond to environmental perception or the *perceived environment* and the *real physical environment*. The former is here understood to mean the environment which is the same for all people, the environment which can be directly perceived on the spot through the senses, not an 'after-image' distorted by experience, memory, etc. The latter means man's true ecological relationship to his environment on which his survival depends. Between the two levels there remains the internal concept of the external world corresponding to environmental cognition or the *cognized environ-*

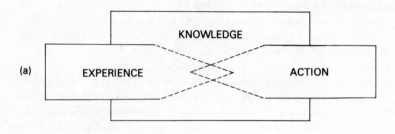

(a)

(b)

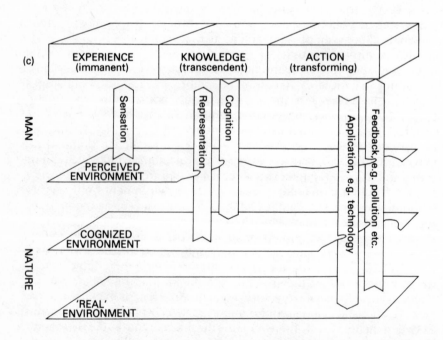

(c)

FIGURE 3.3 Structure and process of geographical enquiry

ment. The representation of the cognized environment constitutes all knowledge of the environment together with the imagination, etc. This 'cognitive map' of the environment is constantly subject to change by all people. It reflects the 'Weltbild' of any one time.

As knowledge has developed so the cognized environment created by mystical, speculative and subsequently rational thinking has been subjected to an increasing number of influences from perception. The development of empirical science meant that attempts were made to identify the cognized environment as nearly as possible with the real environment. The perceived environment was, in turn, looked upon only as an inductive instrument for extracting empirical knowledge about the real environment. The aim was to attain a single objective environment which scholars see as their object of study and which they regard from a distance as outsiders. In this *land-centred* approach the cognized environment represents only a logico-formal level of the real environment, where phenomena are believed to function with the same logic as that in which man thinks. On the other hand, there was also the parallel *man-centred* approach, namely the attempt, either implicitly or explicitly, to keep the perceived environment (on the part of phenomenological empiricists) and the cognized environment (on the part of phenomenologists) presented by means of 'soft' data separate from the real environment, which was presented by means of 'hard' data. The history of geography as a discipline may be seen as a melting pot of these different approaches.

Geographical enquiry has gradually been moulded into the praxis of a specific group, namely geographers. The formation of this group as, in its turn, depended on the institutionalization of geography as a discipline. When this took place, geographers began to act as a social group with all the characteristics typical of such a group (see Bartels, 1970; Hard 1973b; Hurst 1973; Buttimer 1974). The link between social group and factual content originated from the fact that scholars both as individuals and as a group were obliged to make a *choice* of the subject matter to be taught and to be studied and of method. It is *what* is studied that has determined the content of geography. In making his choice, the scholar has permitted the question value to affect him, at least implicitly, through his own or his school's interests. The totality of choices forms the respective framework for each stage in the history of geography. This totality may – because of the subjective factors mentioned – be illogical: 'however logical an individual scholar's work may be methodologically, the object of that work remains illogical when seen in relation to the choices made by others' (Granö, 1963, 15). Seen from the synchronic point of view this is a heterogeneous collection of different choices which, because of the progression of knowledge and the age of the individual geographer, are of 'different ages'. Diachronically, it is possible to reconstruct different phases in the manner of Kuhn, for example. However, the paradigm concept showing the progress of science in general cannot easily be applied to geography: the use of the paradigm cannot be linked with consideration of the development of geography. It has only been used in attempts (e.g. Chorley and Haggett, 1967) to show how mathematical methods used in science have spread to geography. The users of this term, 'new paradigm', wished to indicate that geography was being reborn in all the grandeur of quantitative and systems analytical methods and the past was dismissed (Granö, 1977).

Once geography had become a discipline, the goal of geographical praxis was based on the transfer of knowledge, on the process of informing and learning. Geographers were expected to disseminate knowledge which corresponded to the pragmatic views held by those decision-makers whose efforts had led to the establishment of geography as a discipline. Their view was that geography was a branch of knowledge that provided useful information. In this way a reciprocity between geographical knowledge and society grew up as a result of praxis. Geography took on the form of a learning process which spread through the schools to the general public and gradually to the university. The status and importance of geography came to be based on how useful or interesting was the information it could offer seen from the viewpoint of each recipient.

5 THE DEVELOPMENT OF THE GEOGRAPHICAL DOMAIN IN THE CONTEXT OF INSTITUTIONALIZATION

The institutionalization of geography, as contained within the corresponding process of science in general, and its relationship to geographical knowledge, as part of scientific development in general, is illustrated schematically in table 3.1. The table shows the development since the age of enlightenment. In the first two columns are shown changes in institutional structure. The third shows certain of society's goals, some of which have been related to the development of science. This, in turn, is portrayed by certain successive concepts that indicate streams of thought in science and which have constituted science's 'response' to society's goals. The internal development of geography in this context is portrayed, on the one hand, as goals or aims and, on the other, as geographers' responses to these aims.

The successive labels in the different columns, which have been chosen to represent some characteristic feature, do not by any means cover the whole of the period nor do they give a true picture of the situation in any given country. The purpose of the table is only to show that the development of geography has been a gradual process of change resulting from the formation of an institutionalizing social group, namely geographers, and the choice that this group has made at any one time in its history. The

TABLE 3.1

Contextual development in geography since the age of enlightenment.

Year	Institutional Structure of Geography	Institutional Structure of Science	External Goals of Science	Response of Science	Internal Goals of Geographers	Response of Geographers
Present			Global orientation Quality of life	Interdisciplinary research Environmental Science	Re-identification Social relevance	Social Humanism Perceptionalism Behaviourism
1950	Profession	Projects Research Institutions	Social Policy National prestige Economic growth	Social Science 'Big' Science Specialization	Application Scientification	Quantification
1870	Discipline Societies		Industrialism Neo-Colonialism Free trade	Science + Technology Experimental Science	Identification	Regionalism Environmentalism Exploration
1800	Pre-discipline	Universities Learned Societies	Progress	Romanticism Natural Philosophy	Propaedeutic Education	Naturalism Cataloguing

B

internal goals of geography and the external structures of the environment are reflected in geography's development at different times in
different ways.

As far as the institutionalized structure of science is concerned, the
eighteenth century may be regarded as the period of academies and
learned societies, the nineteenth as the period of the universities and the
twentieth as the period of research institutes. The latest feature has been
the formation of problem-oriented projects independent of discipline
structure.

The entry of science at the turn of the nineteenth century into the
universities, which had hitherto only served society's teaching goals,
made it possible for geography to become a university institution too. The
union of science and teaching resulted in the setting up of different
disciplines. This was quite the opposite of what had occurred during the
time of the academies. Defining the content of knowledge brought about
the setting up of a social body, a staff, for each discipline. The existence of
a discipline structure was a necessary pre-requisite for the birth of a
discipline called geography.

Up to this stage geographical knowledge had only been linked with
teaching in schools and universities and there had been no body of
scholars whose sole occupation was geography. At the beginning of the
seventeenth century the content of geography had gradually begun to
free itself from the grips of theology (Büttner, 1973, 1975); its development reflected the general progress of knowledge even though it had no
disciplinary structure of its own. The inventory-like cataloguing of different countries and peoples remained, however, the main focus of the
teaching of geography, which continued to be looked upon as an ancillary
to history and politics. But the naturalistic trend rooted in the age of
enlightenment and in romanticism also spread to the teaching of geography, and as a result of this trend there was a move towards studies of
nature as man's environment. The fundamental idea underlying this was
the cognitive examination of the environment based on perception as
applied to the pedagogical goals of the times (principally Rousseau and
Pestalozzi). Consequently, for the first time there arose a difference
between cognition of the environment related to the learning process,
which might be called the *geography of nature* or of the *physical world*, and
empirical natural science, from which developed, as far as geography was
concerned, *physical geography*. The geography of nature meant a start was
made on classifying the earth's surface into natural regions. Unlike states,
such regions are not subject to short-term changes, which were rather
frequent in those politically disturbed times. In his history of geography
Wisotzki (1897) united this 'geographia vera' trend, which may be considered to have influenced events from 1726 (Leyser) to 1820 (Wilhelmi),
under the name pure geography ('Reine Geographie'). J. G. Granö

revived this name (1922, 1929) to describe the perceptual geography that he had developed.

The geography-of-nature tradition was peculiar to geography in so far as it led to the first attempt at the turn of the nineteenth century to establish geography as an independent discipline and the important change in its status from just being an ancillary to the study of history and politics. These strivings by the advocates of pure geography were quite in keeping with the spirit of the times for botany, for example, was at the same time trying to cut itself free from medicine. This led to Linné's system, which was based on the characteristics of plants and not on their use in the preparation of medicaments.

Before it had become firmly established as a discipline, geography began the process of institutionalization not just as a unit disseminating teaching but also in the form of scientific societies. These were founded largely in connection with explorations from the 1820's onwards and were modelled on the learned societies of the eighteenth century. These societies did not bestow upon geography the nature of a clear identity; instead, geography was regarded by the societies only as the sum total of work done in different disciplines in the particular area studied, not as a science in itself. The *geography of societies* has, in many countries, maintained this divergent content from *geography as an academic discipline*.

Since Hettner's work (1927, 74–90), the early nineteenth century, the period of Ritter and von Humboldt, has been regarded as the rebirth of geography. It is perhaps more justified, however, to refer to it, as did James (1972, 147), as the close of the old geography since von Humboldt represents the old exploration tradition of the learned societies of the age of enlightenment and Ritter the old university tradition according to which geography was but a propedeutic subject for studies of history and politics.

It is surprising that geography did not achieve the status of an important discipline in Germany in the early 1800s. No geographers had ever gained such a position in science and society as had von Humboldt and Ritter. There was nobody to continue their work on their deaths in 1859 and society made no special attempt to procure someone to carry on their work. An explanation for this might be that at the time the founders of experimental science were trying to gain an entry into the universities: they found their way blocked by professors of natural philosophy and as a result there ensued a struggle. This struggle was eventually won by the scientists. But early geographical thinking about the relationship between man and nature had been traditionally based on natural philosophy. On the other hand, exploration-type geography had gained no position in the universities.

The fact that the Prussian government decided to found chairs of geography in universities in the 1870s arose directly out of the need to educate school teachers in the subject but a fundamental reason for this step was that geographical knowlege was looked upon as beneficial, the reason being that the German nation was undergoing a process of change subsequent to the wars it had been involved in. A further reason was that geography was seen as field of knowledge connected with explorations and this was becoming increasingly important because of the heightened interest in colonialism. But industrialization arising out of the application of science, which was taking place at the same time, did not affect the development of geography, despite its importance as a factor unifying science and technology (Andersson, 1975; Granö, 1976a; Böhme, 1977, 336–8).

It was the external goals of society that brought about the establishment of geography as an academic discipline. This took place without any notable contribution from any other scientists. The late appearance of geography on the university scene, which had already been divided into disciplines, gave rise to many difficulties. The first geographers in Germany and many other countries had received their training in other fields. The transition to geography brought about the strong need for the creation of an identity. Their goal was to attain a respected position among the established disciplines already existing in the scientific community.

External goals did not, however, constitute the guidelines for the development of geography during the first period of its existence as a discipline. Instead, the creation of an independent identity for geographers was a more important target. The content of geography and geographical thought was, at the time, divided into two parts: public thought and geographers' formal thinking. Geography was no longer seen and experienced as part of everyday life: instead it gradually developed into a doctrine. 'Public geography', i.e. pre-scientific thinking (p. 23), continued to live on as some kind of 'sub-science' in spite of the fact that the degree of rationalization was increasing all the time (see Bartels, 1970; Hard, 1973a; Berry, 1973).

An important influence on the choice made by early geographers was the attempt to establish an object of study that could be regarded as geography's own and that differed from that of other disciplines. Consequently, there was introduced into the choice the value that has continued to interest geographers: the 'geographicality' of subjects. There was nothing exceptional in this, for it was a striving common to many other disciplines. As the importance of the experimental sciences grew and the natural philosophical and humanistic tradition correspondingly declined there began to appear a certain exceptionalism in relation to other sciences. This expressed itself in the fact that the unified and monistic geography that its identity required began to appear only in the form of a programmatic formal objective in academic teaching. In practical research work, however, geographers often followed in the footsteps of other specializing sciences. This gradually led to physical geography forming a group made up of different fields with no single theoretical framework. Certain branches of geography were also transferred to neighbouring disciplines. The study of the physical environment, for example, ignored biogeography and became one-sided so that the separation of physical geography from human geography was given a heavy emphasis. The work of Hettner (from 1895) and many others to unify geography both at this time (e.g. Mackinder, 1887; Herbertson, 1905; Banse, 1912; Passarge, 1913; Berg, 1915) and later should be seen in the light of this development.

The identity of geography presupposed on the one hand a unified monistic discipline which, on the other, should have its own concrete object of study in the real world as befitted a science. Environmentalism, as the study of relations, did not fulfil this demand. Perceptual-cognitive studies in the tradition of the geography-of-nature took on the form of objective regionalism à la Hettner, or the region as a holistic unit became the object of study of geography. The basis for the choice of regions as the object of study was provided by the perceived environment, from which has developed the cognized environment (figure 3.3). A number of general concepts characteristic of the environment were adopted in different countries: in Russia 'zone' (Dokuchayev, 1898; Berg, 1915), in Germany 'Landschaft' (Schlüter, 1906; Passarge, 1908, 1913), in the British Empire global 'natural region' (Herbertson, 1905), in France 'pays' (Vidal de la Blache, 1910; Brunhes, 1910), etc. The 'Landschaft' concept, as an intellectual movement in German-speaking areas, was strongest and best suited to the symbol world of geographers (see Hard, 1969, 1970). Attempts were made to eliminate the perceptual basis since it was subjective and instead the aim was to achieve a rational synthesis of the real environment instead of wholeness and immediacy of contact with nature. In fact, the perceptual-cognized environment was preserved in regional geography but adapting it to the same framework as the scientific concept of the real environment continued to present a problem.

Retaining the region as an objective holistic unit was not successful at a time when an essential characteristic of scientific method was the specialization and isolation of the object to be studied from its local connections. The prevailing theory of science at the time (positivism), while uniting all fields of study so that they were studied by means of the same methods, rejected any holistic view.

The regional concept, with man and a differentiated natural environment as geographers' object of study, could not keep geography together and united. At this stage there was a new kind of physics which was spreading to the social studies; this was based on probability and used statistical quantification as its main method. The background to this was the change that was taking place in the relations between society and science. Science was seen as the best way to achieve economic growth. This resulted in geography disparaging its own object of study and instead laying emphasis on quantitative methods. The first geographers to use these methods did not consider the basic ideas of the philosophy of geography nor research policy. This period of 'social physics' was also non-humanistic: man was seen only in terms of statistical distribution and region as a topological surface.

During the identification period the development of geography was isolated from society. This state of affairs continued during the quantification period, when in fact the external goals imposed by society influenced geography through the agency of other sciences. An essential change in structure started at this time and geography came into direct contact with society.

Simultaneously, a fundamental change had taken place in the relation between science in general and society. Science, which in industrialized society's view had been largely only a basis for technology, began to participate in planning the development of the whole of society as part of social policy. Science itself has also become the object of planning. Science is regarded as such a powerful influence on society that conscious attempts are being made to steer its development. We have seen the birth of *science policy*, the empirical part of which might be termed *research policy* (see e.g. Spiegel-Rösing, 1973; Salomon, 1977).

Geography, too, abandoned its passive educational role in which knowledge was produced and disseminated as a way of building up a 'Weltbild', or useful knowledge was created for the use of others. Instead, geography began to contribute actively to a transformed and replanned world. Applied geography was created and geography became a *profession* outside the small world of the university.

This structural change also affected the content of geography. The limited concept of man and his environment which was characteristic of the scientification period and its stress on methods was not sufficient in this new structure. In addition, there came behaviouristic and perceptual geography, in which man as a cognitive individual plays a more important role. The stress on perceptual geography has not resulted, as was earlier the case, from the observation of the environment as such. On the contrary, perceptual geography was studied as a way of explaining human behaviour. The basis of this was not, as in the case of deterministic and possibilistic environmentalism, the real environment but rather the perceptual-cognized environment (figure 3.3).

With the tendency for society's external goals to be transferred away from economic growth and more and more towards problems of environmental protection and regulation there has been a strengthening of ecology and the environmental sciences. This presupposes an integrated study of man and his real environment and attempts have been made to solve the problem using an interdisciplinary approach. The results achieved have not always measured up to expectations because scholars from specialized disciplines have continued, even when working in teams, to pursue the research methods of their own disciplines.

The consequences of this are to be seen in geography in the form of an orientation towards attempts once more to integrate man and the natural environment. Although quantitative spatialism and general systems theory have been applied to both physical and human phenomena, these methods have not yet produced a practicable research programme for a new regional geography based on the integration of man and the natural environment. Attempts have, however, been made to reverse the scientific method's process of isolation of the object of study and return it to its real local and temporal context and situation.

In this approach, the object of study is a combination of empirical data from nature and human life in the outside observed real environment. The rational theories based on the juxtaposition of these data may be regarded as constituting the scholar's logico-formal concept of the real environment (p.25). But in perceptual and behaviouristic geography environmental cognition represents imaginary concepts of experience, and the observer and the observed environment become intertwined. The result of this is a psychological view of the environment. If, however, the study of the real and perceptual-cognized environment is integrated, the result is a 'cross-science' of physics and psychology dealing neither with mind nor matter, but with both together in some way. Many philosophers (e.g. Nicod, 1924; Russell, 1914, 1935; Kaila, 1941, 1960) have given consideration to this problem. The history of geography, beginning with the geography-of-nature tradition, shows that this question has been geography's concern, albeit often only implicitly.

The traditional integration of geography (regional geography) had achieved its position in a process of identification based on its own internal programmes and not as a result of the influence of society's external goals. The present trends towards integration, however, conform more to society's demands.

The new humanistic geography and radical geography, which have been united in table 3.1 under the name of 'social humanism', have developed the perceptual-cognitive approach further to a stage where the scholar is identified with his object of study, individuals' or social groups' perceptual-cognized environment. The application of geography is in that case directed, not towards the real environment as expressed in empirical data, nor to the perceived environment as expressed in sense data, nor to the cognized environment as expressed in imaginary concepts or rational theories, but to the future *potential environment as expressed in values.*

Development has led to a situation where geographers are now confronted with a period of re-identification. This does not, however, depend on internal factors within geography but on how the goals and praxis of geographers can be adapted to the future of university structure. Geography has indeed become a profession as a result of its application but it has been able to maintain its strongest existence in the discipline-oriented world of the university. But here, too, the decision-makers have a certain application as their basis, namely the pedagogical application. The future existence and nature of geography is tied up with the future development of university structure, in other words, with the question of the degree to which the new *problem-oriented* trend in society will replace the *discipline-oriented* structure. In practice, this depends on the possibility of developing a problem-oriented type of teaching in the universities. If this can be achieved, then the result will be a restructuring of science that will completely revolutionize the old division based on disciplines.

Acknowledgements

The author gratefully acknowledges valuable comments on this paper from Anne Buttimer, Torsten Hägerstrand and David Seamon.

References

Andersson, G. 1975: *Vetenskapens nytta och frihet* (The freedom and utility of research). English summary. Reports from the department of Theory of Science, University of Göteborg 75.

Banse, E. 1912: Geographie. *Petermanns Geographische Mitteilungen* 58, 1–4, 69–74, 128–31.

Bartels, D. 1970: Zwischen Theorie und Metatheorie. *Geographische Rundschau* 22, 451–7. English translation: Between theory and metatheory. In Chorley, R. J., editor, *Directions in Geography* (London, 1973) 23–42.

Beck, H. 1954: Methoden und Aufgaben der Geschichte der Geographie (Methods and tasks of the history of geography). English summary. *Erdkunde* 8, 51–7.

Berg, L. S. 1915: Predmet i zadachi geografii (The subject and scope of geography). *Izvestiya Russkogo Geograficheskogo Obshchestva* 51, 463 ff.

Berger, H. 1887: *Geschichte der wissenschaftlichen Erdkunde der Griechen* (Leipzig).

Bernal, J. D. 1939: *The social function of science* (London).

Berry, B. J. L. 1973: A paradigm for modern geography. In Chorley, R. J., editor, *Directions in Geography* (London).

Böhme, G. 1977: Models for the development of science. In Spiegel-Rösing, I. and de Solla Price, D., editors, *Science, Technology and Society* (London) 319–51.

Brunhes, J. 1910: *La géographie humaine* (Paris).

Buttimer, A. 1974: Values in geography. *Resource Paper, Association of American Geographers*, 24.

Buttimer, A. 1976: Grasping the dynamism of lifeworld. *Annals of the Association of American Geographers* 66, 277–92.

Büttner, M. 1973: *Die Geographie Generalis vor Varenius: geograpischts Weltbild und Providentiallehre.* (Erdwissenschaftliche Forschung 7, Wiesbaden).
Büttner, M. 1975: Kant und die Überwindung der physikotheologischen Betrachtung der geographisch-kosmologischen Fakten (Kant and the surmounting of the physico-theological interpretation of geographic-cosmological reality). English summary. *Erdkunde* 29, 162–6.
Chorley, R. J., and Haggett, P. 1967: Models, paradigms and the new geography. In Chorley, R. J. and Haggett, P., editors, *Models in Geography*, 19–42 (London).
Dokuchayev, V. V. 1898: *Kuchschenie o zonah prirody* (The study of natural zones) (St. Petersburg).
Granö, J. G. 1922: *Die landschaftlichen Einheiten Estlands* (Tartu, Loodus).
Granö, J. G. 1929: Reine Geographie. *Acta Geographica* 2 (Helsinki).
Granö, O. 1963: Maantieteen asema tieteellisessä tutkimuksessa (The position of geography in scientific research). *Valvoja* 83, 11–17.
Granö, O. 1976a: Vetenskapens utveckling och forskningspolitik som bakgrund för geografins roll (The development of science and science policy and their importance for geography). English summary. *Svensk Geografisk Årsbok* 52, 33–45.
Granö, O. 1976b: *Man's dualism in relation to his environment.* Working Group on Man-Environment Theory, Working Note 8 (International Geographical Union).
Granö, O. 1977: Geography and the problem of the development of science. *Terra* 89, 1–9 (Finnish with English summary).
Gustavsson, S. 1971: *Debatten om forskningen och samhället* (The debate on research and society). English summary. *Skrifter utgivna af Statsvetenskapliga Föreningen i Uppsala* 54.
Hard, G. 1969: Die Diffusion der 'Idee der Landschaft' (The diffusion of the idea of the landscape). English summary. *Erdkunde* 23, 249–64.
Hard, G. 1970: Die 'Landschaft' der Sprache und die 'Landschaft' der Geographen. *Colloquium Geographicum* 11, 1–278.
Hard, G. 1973a: *Die Geographie* (Berlin).
Hard, G. 1973b: Die Methodologie und die 'eigentliche Arbeit' (Methodology and 'true work'). English summary. *Die Erde* 104, 104–31.
Herbertson, A. J. 1905: The major natural regions. *Geographical Journal* 25, 300–12.
Hettner, A. 1895: Geographische Forschung und Bildung. *Geographische Zeitschrift* 1, 1–19.
Hettner, A. 1927: *Die Geographie* (Breslau).
Hurst, M. E. E. 1973: Establishment geography. *Antipode* 5, 40–60.
James, P. E. 1972: *All possible worlds* (New York).
Kaila, E. 1941: Über den physikalischen Realitätsbegriff. *Acta Philosophica Fennica* 4 (Helsinki).
Kaila, E. 1960: The perceptual and conceptual components of everyday experience. Translation in progress. Finnish in *Ajatus* 23, 50–115.
Kuhn, T. S. 1962: *The structure of scientific revolutions* (Chicago).
Lenntorp, B. 1976: *Paths in space-time environments. Meddelanden från Lunds Universitets Geografiska Institution, Avhandlingar* 78 (Lund).
Lüdde, J. G. 1849: *Die Geschichte der Methodologie der Erdkunde* (Leipzig).
Mackinder, H. J. 1887: On the scope and methods of geography. *Proceedings of the Royal Geographical Society* 9, 141–74.
MacLeod, R. 1977: Changing perspectives in the social history of science. In

Spiegel-Rösing, I. and de Solla Price, D., editors, *Science, technology and society* (London), 149–96.

May, J. A. 1970: *Kant's concept of geography and its relation to recent geographical thought*. University of Toronto Department of Geography Research Publications 4 (Toronto).

Merton, R. K. 1938: Science, technology and society in seventeenth-century England. *Osiris* 4 (Reprinted New York 1970).

Mulkay, M. J. 1977: Sociology of the scientific research community. In Spiegel-Rösing, and de Solla Price, D., editors, *Science, technology and society* (London), 93–148.

Nicod, J. 1924: *La géometrie dans le monde sensible* (Paris).

Paassen, C. van 1957: *The classical tradition of geography* (Groningen).

Paassen, C. van 1976: Human geography in terms of existential anthropology. *Tijdschrift voor Economische en Sociale Geografie* 67, 324–41.

Passarge, S. 1908: Die natürlichen Landschaften Afrikas. *Petermanns Geographische Mitteilungen* 54, 147–60, 182–8.

Passarge, S. 1913: Physiographie und Vergleichende Landschaftsgeographie. *Mitteilungen der Geographischen Gesellschaft in Hamburg* 27.

Peschel, O. 1865: *Geschichte der Erdkunde bis auf Carl Ritter und A. v. Humboldt* (Munich).

Radnitzky, G. 1974: Toward a system philosophy of scientific research. *Philosophy of the Social Sciences* 4, 369–98.

Relph, E. 1970: An inquiry into the relations between phenomenology and geography. *Canadian Geographer* 14, 193–201.

Relph, E. 1976: *Place and placelessness. Research in Planning and Design* 1 (London).

Richthofen, F. von 1883: *Aufgaben und Methoden der heutigen Geographie* (Leipzig: inaugural lecture).

Russel, B. 1914: *Our knowledge of the external world* (London).

Russel, B. 1935: *Religion and science* (London).

Salomon, J. J. 1977: Science policy studies and the development of science policy. In Spiegel-Rösing, I. and de Solla Price, D., editors, *Science, technology and society* (London), 43–70.

Samuels, M. S. 1971: *Science and geography: an existential appraisal* Ph.D. dissertation, University of Washington, Seattle.

Schlüter, O. 1906: *Der Ziele der Geographie des Menschen* (Munich and Berlin).

Spiegel-Rösing, I. S. 1973: *Wissenschaftsentwicklung und Wissenschaftssteuerung* (Frankfurt/Main).

Spiegel-Rösing, I. 1977: The study of science, technology and society (SSTS): recent trends and future challenges. In Spiegel-Rösing, I. and de Solla Price, D., editors, *Science, technology and society* (London) 7–42.

Tuan, Yi-Fu 1971: Geography, phenomenology and the study of human nature. *Canadian Geographer* 15, 181–92.

Tuan, Yi-Fu 1974: *Topophilia* (Englewood Cliffs, N. J.).

Vidal de la Blache, P. 1910: Régions françaises. *Revue de Paris*, 821–49.

Vidal de la Blache, P. 1913: Les caractères distinctifs de la géographie. *Annales de Géographie* 22.

Wisotzki, E. 1897: *Zeitströmungen in der Geographie* (Leipzig).

4

Institutionalization of Geography and Strategies of Change

Horacio Capel

One of the characteristic traits of current social science is the generalized awareness of crisis and, in particular, the crisis of existing disciplinary divisions.[1] The unsatisfactory character of the acknowledged limits between the distinct scientific specialities is revealed through the efforts made within each discipline to surpass these limits and incorporate theories, methods or points of view derived from neighbouring disciplines, thus establishing connections that until recently were totally unsuspected. The recent evolution of our science after the discovery of the wide field of perception and behavioural geography, and the establishment of connections between geography and psychology are examples of the above, among others which could be cited.

It is increasingly evident that a reorganization of the fields of knowledge and of the limits between distinct specialized branches of the science is taking place. Some scientists believe that it is necessary to combine many existing branches of science with a view towards a reorganization of the disciplinary fields, to make possible a freer, more imaginative scientific viewpoint in the solution of concrete problems.

In such a situation, the study of the process of institutionalization of the sciences may prove particularly fruitful. If we can come to understand the factors that lead to the institutionalization of some sciences and the failure of the embryonic sciences that could facilitate alternative frameworks of scientific development, perhaps we would be able to understand their later evolution and proceed more easily to a reorganization of the fields of knowledge.

From this perspective, an investigation of the process of institutionalization of geography, and of the appearance of the scientific community of geographers, is particularly interesting. But the analysis of the origins and evolution of this science and the history of geographic thought must not be done apologetically – as is common in the histories of geography and, in fact, in the greater part of the histories of the different sciences – but rather set forth general problems relevant to the distinct social sciences, to which geographers can contribute the knowledge and experience of their own science.

The essential argument of this paper can be summarized thus: despite the antiquity of the term 'geography', the geography of the twentieth century has little to do with that of the nineteenth. Today's geography has its origins in the process of institutionalization that from the mid nineteenth century, and after a period of decline, leads to the appearance of the scientific community of geographers, extending without interruption to the present time. The factors which lead to the institutionalized existence of this community are directly related to the presence of geography in primary and secondary education at the time when the European countries begin the rapid process of diffusion of elementary education; the necessity to train geography teachers for primary and middle schools was the essential factor which led to the institutionalization of geography in the university and the appearance of the scientific community of geographers.

The emergence of the community depended upon the support of governments, geographical societies and of some scientists, and met with violent opposition from a good number of scientists from other fields. In the struggle for recognition, the members of the new community (whom we shall call geographers) had to strive to demonstrate the specific character of their science, arguing its objective and defining its limits with respect to the sciences practised by other scientific communities (geologists, historians, ethnographers, ecologists, sociologists, etc.). Later, the growth of the community and/or the blockage of traditional teaching careers – as a result of, for example, the competition of naturalists and historians in secondary education – led to the search for new professional opportunities for the members of the community; this influenced the appearance of new directions in the discipline and their corresponding theoretical justifications. The members of the scientific community of geographers possessed, thus, some specific interests and objectives; as a result some very coherent and permanent strategies can be found in the scientific texts produced by the community. This paper will demonstrate some of these strategies, analysing the process of institutionalization and the theoretical writings of some national communities of geographers. The necessity to introduce the perspective of the sociology of science in the study of the history of geographical thought is defended as a complement to other, now more usual approaches.

THE DECLINE OF A SCIENCE

The most specifically geographical texts produced during the first two centuries of the Modern age were either geographies or cosmographies that followed the Ptolemaic conception and constituted, in essence, a localization of toponyms for the use of navigators and astronomers, or descriptive geographies that contained narrations on the characteristics of the lands, customs and social organization of different countries, and that responded to the interest and curiosity of a wide public.

The *Geographia Generalis* of Varenius (1650) represented an attempt to develop a general geography that would permit the subject to be considered a science, and would facilitate the later development of regional or special studies. For Varenius, this general geography, which 'considers the earth in its entirety', was solely physical, mathematical and astronomical, while the human phenomena, which he affirmed as 'less pertinent to geography', entered into consideration only in the study of regions, and only then because, as he wrote, 'something must be conceded to the habit and usefulness of those that study geography' (Capel, 1974).

During the Modern age there frequently existed an instrumental and practical relationship between history and geography, particularly in their chorographical aspects. Geography and chronology were considered the basic columns of history, so that geography became an auxiliary science of history. This union between geography and history, based upon tradition and upon practical considerations, was elevated to the theoretical level towards the end of the eighteenth century by Kant, who considered that the two sciences are separated in the scientific system because they study unique phenomena in time, or in space. 'History and Geography', writes Kant, 'could be denominated thus, as description, with the difference that the first is a description in time and the second a description in space.' (Schaefer, 1953).

In the European centres of higher education, the relationship between geography and history was close until well into the nineteenth century, and on many occasions geography was considered a simple auxiliary of history, as demonstrated by the existence of professors of geography and chronology in some universities as late as the middle of the century. And, in a way, it is to this relationship that is owed the geography of K. Ritter, considered one of the fathers of modern geographical science. When the University of Berlin, founded in 1810, called on Ritter as professor of geography, he still could not decide between the one subject and the other (having been professor of history at Frankfurt). Ritter was, of course, a geographer through his university activities as such, but his interests lie as much in history as in geography.

A case very distinct from Ritter's is that of A. von Humboldt, who can only with difficulty be considered a geographer in the strict sense of the word. Undoubtedly he consciously tried to create an earth science, the physics of the globe, in which he was influenced by the ideas about the unity of nature dominant in the Germany of his youth and particularly by the thought of Goethe (Beck, 1959–61). It is this unitary conception which led him to make observations in such diverse fields and which, without doubt, is at the root of his voyage to America. He considered this to be something more than a scientific expedition with first rate astronomical instruments; as he says in a letter written the day of his departure for America:

> All of this is not, however, the principal objective of my journey. My eyes must always be fixed upon the combined action of the forces, the influences of inanimate creation upon the animal and vegetal world, upon this harmony. (Letter to Von Moll, 5 June 1799, cited by Minguet, 1969, 61)

It is an idea that he maintained all his life, and that motivated the production of such different works as the 'Historical relation of the voyage to the equatorial regions of the new continent' (1818) – where he affirms that

> the major problem of the Physics of the world is to determine . . . the external bonds that connect the life phenomena to those of inanimate nature

– or the great work of his maturity, the *Cosmos*. Hence came the variety and richness of his observations, from geomagnetism to archaeology, and thus his work may be viewed as the foundation of diverse specialized sciences such as botany, meteorology, and geophysics. Humboldt was, above all, a naturalist of a rigorous scientific spirit and a defender of the empirical method, but imbued with the natural philosophy and the romantic ideas of his era. His work was valued by all interested in the earth sciences, and by 'geographers' of the epoch, essentially through geographical societies with a naturalist membership of botanists and geologists such as Murchison or Somerville.

Despite the existence of such a figure as Ritter, numerous accounts show that in all Europe, geography was a science in profound crisis and even upon the point of disappearance during the first half of the nineteenth century. This occurred as much in its academic status as in its popular esteem.

1. In Germany, the historians of geography habitually point out that the ten years that followed the death of Humboldt was a decade of crisis.

Kretschmer (1930) has written that after the death of Humboldt in 1859 came 'a period of paralysation': 'Grand personalities did not exist, nor were there, properly speaking, schools; the seventh decade of the century forms a gap in the constant development of geography, a vacuum only interrupted by a few sensational discoveries.' This point of view is shared by Hettner (1898) and following him, Hartshorne (1959, 29). A similar opinion is held by Paul Claval (1974), alluding explicitly to the crisis of 1860–70, and by G. R. Crone (1970, 32) who indicates that 'after the death of Ritter, the impetus given by him gradually died away and for a time geographical theory received little attention in Germany.'

2. In France during the first two thirds of the nineteenth century, geography passed through a profound crisis, contrasting with the great development of the science in the preceding centuries. Contrary to the high esteem it enjoyed during the eighteenth century, and contrary to the awareness of its usefulness for navigation and commerce – translated into schemes such as the creation of a Geographical Museum during the Revolution (Broc, 1974) – the nineteenth century represents a 'long purgatory' which 'is aided by a recess of curiosity, a true regression' (Broc, 1974, 25). The accounts of this regression are varied, and some have been put together by Numa Broc (1976, 227) who insists that 'in the first half of the nineteenth century, geography does not enjoy high consideration', and demonstrates how the best-known geographers of the epoch, such as Malte-Brun and Balbi, scientists and public authorities considered geography a 'descriptive science', 'of facts and not of speculation', 'of the exclusive dominion of the memory', a science, in short, 'that must not try to go back to the cause and to explain', because 'although these be speculations of great importance, they are outside the domain of geography' (Broc, 1976, 227–8). After analysing these accounts, Broc concludes that for most of the authors cited, 'geography is not a science . . . but simply a practical discipline, appropriate for supplying the diplomat, the soldier, the merchant, with immediately useful information. In the schools it was considered no more than one of the "two eyes of history", beside chronology' (Broc, 229). Proof also of this slight esteem is the fact that in the logical classifications of science from the end of the eighteenth century, geography disappears or is found divided: 'nowhere appears the idea that the characteristic of geography could be, precisely, to integrate the physical and political facts' (Broc, 1976, 230).

In fact geography was studied alongside history, and provided the background necessary to understand historical events, principally of classical history. Thus it happened that the only university professorship in geography that existed at the Sorbonne from 1809 (under the title Geography and History till 1812) was occupied by historians, who according to L. Dussieux in 1883, confined themselves 'with a pedantic solemnity to the geography of Homer and of Herodotus' (cited by Broc, 1976, 225).

At the popular level geography had some impact through the publication of geographical encyclopaedias and informative magazines for the general public. The work of the Danish Malte-Brun, *Précis de Géographie Universelle* (1810–29), constituted, in this sense, the fundamental geographical work for many years. The curiosity of the general public for the exotic countries reflects itself above all in enterprises of illustrated journalism like the *Journal des Voyages* created by Malte-Brun in 1808 (Meynier, 1969), or *La Tour du Monde* or the *Lectures Géographiques*, to cite a few of the more expressive titles.

The works of Humboldt and Ritter did not have immediate impact, despite the fact that the former lived for a time in Paris, published much of his research in French, maintained throughout his life an intense exchange of letters with French scientists, especially with Arago, and influenced the creation of the Geographical Society of Paris (1821). One can, however, indicate one work comparable to that of Humboldt's *Cosmos* in Eugène Cortambert's *Physiographie: Description générale de la nature pour servir d'introduction aux sciences géographiques* (1836), in which the author tried to combine 'the general notions isolated in the special treatises of geology, geography, botany and meteorology.' With this work, Cortambert anticipated ideas of the *Cosmos* (Freeman, 1961, 32), although in a lesser way, permitting Broc to qualify this work as a 'poor man's *Cosmos*' (Broc, 1976, 229).

3. In Great Britain geography also seems to enter into crisis during the first half of the nineteenth century, above all from the fourth decade on. At the University of Oxford, geography had been taught from the sixteenth century to the eighteenth, 'but the changes in the educational system of the university led to the decline of geography in the nineteenth century' (Scargill, 1976, 440). In fact, in the first half of the century, courses in mathematical geography were given at only one college 'for a small audience' and among historians 'geography hardly received better treatment'. The reorganization of the course outlines in 1850 made possible the reconsideration of geography, theoretically, now that it had 'confirmed itself in its old place in classical studies and was made reference to in the regulations for the natural sciences (in the form of physical geography) and for modern history'(Scargill, 1976, 440). But the professorships remained unoccupied, and geography continued undeveloped.

In London, in 1833, a professorship in geography at University College was established, awarded to Captain Maconochie, then secretary of the Royal Geographical Society. But the post was occupied for only a short time, as its holder left for overseas, and it remained vacant because geography 'does not seem to be considered a part of general education. Not even the acknowledged distinction of the late Professor could obtain a numerous class' (cited by Crone, 1970, 28).

During the first half of the nineteenth century, the distinct branches of the natural sciences developed so much more rapidly than geography, that the latter, particularly physical geography, came to be considered part of one of these disciplines. Very significant, in this respect, is the change in section 'C' of the British Association, which was originally entitled 'Geography and Geology' but was changed to 'Geology and Physical Geography' in 1839. Later, however, in 1851, the section 'Geography and Ethnology' was created and survived until 1878 (Baker, 1948; cited by Freeman, 1961, 38).

The close relationship between physical geography and geology is revealed in the work of R. I. Murchison (1792–1871), a geologist of international repute and director of the Museum of Applied Geology, who was one of the founders of the Royal Geographical Society and later its president. Throughout his life, Murchison considered physical geography and geology as two 'inseparable scientific twins' and held a view of geography that would hardly be shared by today's community of geographers: on one hand he valued above all the *Cosmos* of Humboldt and on the other, considered Ritter rather as a historian (Crone, 1970, 30).

The need to establish relationships between separate phenomena, as shown by Humboldt in Germany, appears explicitly in the work of Mary Somerville (1780–1872), considered 'the first English geographer, by publishing the first textbook on physical geography written in English' (Sanderson, 1974, 410). In 1836 she published a work on the connection between the different physical sciences, and in 1848 her famous *Physical Geography*, which had been through seven editions by 1877. The work is inspired by the same spirit as Humboldt's, as shown by Somerville's use of Bacon's aphorism: 'no natural phenomena may be studied by itself, but, to be understood, must be considered in connection with all of nature.' The same point is demonstrated by her definition of physical geography as 'the description of the earth, the sea and the air, with its animal and vegetable inhabitants, of the distribution of these organized beings and the causes of this distribution' (cited by Sanderson, 1974). The inclusion of man in works of physical geography was a common practice in the epoch and persisted occasionally until the end of the nineteenth century, although 'there was no a clear conception of what a 'geography of man' should embrace apart from the "distributions" ' (Dickinson, 1976, 12).

4. The situation of geography in Italy during the first half of the nineteenth century was not very brilliant either. Geography was, in fact,

during these years, simply an adjunct to history. Towards the end of the century C. Bertacchi alludes clearly to this fact, writing of:

> a time in which geography was considered almost a dependency of history, and did not have value other than as illustration of the historical facts. Thus, as Chronology demonstrates the distribution of facts in time, Geography, which we will call more specifically historical, demonstrates their distribution in space. Yesterday's geography was, before all, an historical geography, that extended to the present has converted itself into political and statistical geography. (Bertacchi, 1892, 571)

The names which the Italian geographical histories customarily cite as principal figures of this epoch (Adriano Balbi, Emanuelle Repetti, Marmochi, De la Luca, Zucagni Orlandini) were not really specialized geographers, but authors who could belong to various social sciences. This shows that in mid-century Italy 'geography does not appear well individualized as a science' (Almagìa, 1961, 419), while Pracchi (1964, 575) considers that if one examines publications in this country between 1800 and 1890, one observes that 'during the greater part of the century, a vague and uncertain concept of geography dominated. . . . always intertwining itself with history, statistics or with other disciplines.'

Among the Italian scientists of the epoch, geography does not seem to have been held in high esteem. Thus, although in the congress of Italian scientists held in Milan in 1844 an autonomous section for geography was created, through the influence of A. Balbi, this autonomy 'disappeared in successive congresses, and geography was again associated with other sciences, above all those of the statistical-historical type or simply humanistic'. This explains why the Casati law of 1859 placed geography among the disciplines of the Faculty of Arts (Almagìa, 1961, 420). Everything seems to indicate that during the greater part of the nineteenth century, geography in Italy was for the most part a little valued subject of 'Arts'. As late as 1892, C. Bertacchi could give himself the objective of 'stripping this science of the character of literary exercise that until now it has been given in most of our schools', since 'a geography like this has no reason to exist'; and he could refer to the case of 'a well known personality who had declared jokingly that he did not "believe" in geography' or that of a politician who considered geography 'among the useless literaries' (Bertacchi, 1892, 572).

5. Russian geography had been forged during the eighteenth century through the expeditions organized for the study of the resources of an immense, little known country. In this sense, the expansionist politics of the Czars during the eighteenth and nineteenth centuries constituted an

essential factor that led to the organization of a systematic plan of exploration. M. V. Lomonosov, director from 1758 of the department of geography of the Academy of Science of St. Petersburg, worked on this project in the last year of his life, and it was according to his plan that the important expeditions of the Academy (1768–74) were organized. The department of geography of the Academy played an important role in the collection of information and cartographic systematization and elaboration until the 1860s; but from this date, the creation of specialized organizations (like the official land registry, the cartographic services of the army and navy) made it lose importance. It was abolished in 1800, at the same time as were created sections of the social sciences (statistics, political economy, history). The Academy of Science continued organizing scientific expeditions during the nineteenth century, which contributed greatly to knowledge of the Russian territory and the development of the earth sciences, but it was the more specialized scientists, not geographers, who played the essential roles.

From mid-century the Russian Geographical Society, founded in 1845, actively contributed to the organization of expeditions, as well as to the study of theoretical questions, man-nature relations (a theme accredited to K. M. Behr) and the study and fixing of the geographical terminology (Valskaya, 1976, 50). It was also important as a vehicle of new ideas through exchanges and discussions with other European centres and scientists (Matveyeva, 1976).

Despite all this, during the first half of the nineteenth century geographers as such did not exist in Russia (Sukhova, 1976, 64), and it was other specialists (geologists, biologists, botanists) who made the most important contributions which today are collected in the geographical histories and who, in particular, studied the problem of the relations between natural phenomena (Sukhova, 1976).

It was also these 'naturalist specialists' who accepted and utilized in the first place, from 1860, the ideas of Darwin (like those of Lyell before him) and contributed to the adoption of a new perspective for the study of the interdependencies between phenomena upon the surface of the earth. In particular, the biologists (botanists and zoologists) concentrated upon the study of interaction and also 'made attempts at biogeographical and zoogeographical spatial divisions, using such terms 'when such a notion did not exist in geography' (Sukhova, 1976, 64), smoothing the way for the development of the work and ideas of another naturalist, V. V. Dokuchayev, on genetic pedology.

With regard to human and economic geography, during the nineteenth century 'it did not exist' (Gerasimov, 1976, 103). Their subject matter was studied in statistics – from 1804, departments of geography and statistics existed at the university according to Gerasimov (1976, 91) – and towards the end of the century in economics and political economy. In fact, it could be said that 'the military-statistical descriptions made by the General Staff were perhaps the nearest things to the economic-geographical investigations at that time' (Gerasimov, 1976, 103).

These accounts all demonstrate the profound decline of geographical science during the first half of the nineteenth century. In them, geography frequently appears as a science on the point of disappearance, lacking academic interest – except for historians – forgotten or impugned by many scientists and appreciated solely by the general public because of the descriptions of exotic countries that it contained.

The decline of geography is comprehensible if we take into account that the first half of the nineteenth century is a time in which a series of sciences that deepened the study of our planet were emerging. What Humboldt called the 'Physics of the globe' or 'Theory of the earth' was developing into a series of specialized scientific branches and began a process of institutionalization and professionalization. The economic transformations experienced in Europe to the middle of the eighteenth century, and especially the industrial revolution, pushed the earth sciences in new directions and provoked unprecedented scientific development. While in earlier epochs, maritime commerce had stimulated the development of science, the needs of industry now promoted diverse scientific branches. The exploitation of new mineral resources and the opening of new means of terrestrial transport (canals, roads, and railways) strongly promoted the development of geology, particularly in the period 1775–1825, when geology began to be considered worthy of specialized scientific study, and thus appeared the profession of geologist (Hall, 1976, 211, 232). The publication of Lyell's *Principles of Geology* (1830–36) was a decisive step in the development of geology, Lyell's definition of which doubtlessly affected geography, since it occupied a field which geography could pretend – and in fact, later claimed – to occupy. Geology is defined as a

> science which investigates the successive changes that have taken place in the organic and inorganic kingdoms of nature; it enquires into the causes of these changes and the influence which they have exerted in modifying the surface and internal structure of our planet. (Lyell, 1830; cited by Hall, 1976, 234)

In a parallel direction, geophysics began to emerge from Physics and the study of geomagnetism received a strong impulse from the development of maritime navigation and the introduction of iron-clad ships. At the same time there developed international cooperation – for example, the creation of the Magnetic Union of Göttingen, organized by F. Gauss in 1834, at the instigation of A. von Humboldt – and the profession of geophysicist, operating the network of observatories created from 1840 (Hall, 1976, 252).

Cartographic studies became a specialized scientific field because of their importance in exploration for resources and in navigation; geodesy also received a strong impulse, and the exact determinations of gravity and curvature became a basic prerequisite for adequate mapping. Mineralogical investigations led physicists and geologists to lively discussion of the problem of the composition of the earth's interior (Hall, 1976, 262). The organization of national networks of meteorological stations gave a clear impetus to the study of meteorology and climatology, but one made by specialists, not geographers. In another related direction, botanists and zoologists were modifying their old taxonomies before the appearance of new ideas that would result in the Darwinian revolution. And an equally rapid evolution is found in the field of the sciences of man.

In this emergent phase of the specialized scientific branches, the problems of articulation of distinct knowledge ceased when the unitary conception of nature with its idealist roots entered into crisis. Thus, as Dickinson has indicated (following G. R. Crone):

> the study of the earth's surface forms as an integrated field was divided among the various sciences that sought to examine processes by reference to individual categories of earthbound phenomena. The concept of natural unities (*Länder*) together with their people, as envisaged by Forster and Pallas and developed by Humboldt and Ritter, was neglected by the scientists. (Dickinson, 1969, 52; Crone, 1951, 1970, 32).

These are the circumstances that explain the profound crisis of geography during the first half of the nineteenth century, a crisis that, despite some suggestions otherwise, also seems to be shown, as we will see, by the evolution of the geographical societies.

THE INSTITUTIONALIZATION OF THE 'NEW GEOGRAPHY'

In contrast to the sombre panorama presented towards the mid nineteenth century, fifty years later geography appears extraordinarily vigorous and expansive in many countries. It is taught in a great number of universities and is present in all primary and secondary education programmes; it receives theoretical contributions from an active community of scientists who edit specialized journals and attend national and international conferences, and are consciously practising a 'new geography'.

The contrast between these situations is striking, and raises questions about the factors which led to the institutionalization and development of a science which only fifty or sixty years before was scarcely appreciated, and about the measure of continuity between this 'new' geography and the 'old' science.

The diffusion of education during the nineteenth century

The essential factor that leads to the institutionalization of geography and the appearance of the scientific community of geographers is the presence of this science in primary and secondary education by the middle of the nineteenth century. The tradition of teaching children elementary notions about our planet through 'geography', and the old relationship between geography and history, probably contributed to the inclusion of the course 'geography' in primary and secondary education pro-grammes, in a residual fashion and generally united with history, at the beginning of the process of diffusion of elementary education all over Europe.

During the nineteenth century, once the debates, frequent at the end of the eighteenth and early nineteenth centuries, on the advisability of teaching the 'inferior classes' were over (Cipolla, 1969), elementary education began to spread and its spectacular results constituted one of the most important social advances of contemporary history. Around 1850, half of the adult European population did not know how to read (60 per cent if one includes Russia), and another 25 or 30 per cent read badly or did not understand what they read; one hundred years later, the proportion of illiterate adults in Europe had been reduced to less than ten per cent (Cipolla, 1969).

The growth of the number of schoolchildren was considerable, as much in absolute terms as in relative. Many European countries doubled the number in the second half of the century. From mid-century to 1910, the number of schoolchildren grew by some 5 million in Germany, 4 million in Great Britain, 2.5 million in France and Italy and close to 1 million in Spain (see table 4.1).

This growth of schooling required a parallel increase in the number of teachers in public and private institutions (table 4.2). Some countries, such as Germany and France, made considerable progress in this direction in a very short time. In Germany, it was prohibited by law to give work to children younger than nine years who had not attended school for at least three years (Cipolla, 1969), and this greatly affected the creation of public schools. Hence it was in Germany (and in Switzerland) that there first appeared a general interest in educational questions. In France, a notable effort had been made in secondary education during the Napoleonic era, although only in 1833 was obligatory basic education established, charged to the municipalities with aid from the state. But it was not until the defeat of 1870 that a concern for the reorganization of

TABLE 4.1

Extension of primary and secondary education in various European countries (number of students in thousands).

	Primary education				Secondary education			
	Germany	*France*	*Italy*	*Great Britain*	*Germany*	*France*	*Italy*	*Great Britain*
1850	—	3 322	—	2 100	—	47.9	—	—
1860	—	4 437 (1865)	1 009 (1861)	—	—	55.9	15.8 (1861)	—
1870	4 100 (Prussia)	4 610 (1875)	1 605	3 100	—	73.9 (1875)	23.2	—
1880	—	5 049	2 003	3 274	—	86.8	32.5	—
1890	—	5 594	2 419	4 288	—	90.8	63.5	—
1900	8 966	5 526	2 708	5 387	—	98.7	91.6	—
1910	10 310	5 655	3 309	6 114	1 016	126.0	164.0	180.5

Source: Mitchell, B. R. 1975: *European historical statistics 1750–1950* (London), 749–70.
For figures in italics, the source is Cipolla, C. 1969: *Literacy and development in the West* (Harmondsworth), table 12.

TABLE 4.2

Extension of primary education in various European countries (number of teachers in thousands).

	Germany	France	Italy	Great Britain
1850		60.0 (1837)		
1860		109.0 (1863)	28.2	7.6 (not Scotland)
1870			41.0	16.8
1880		123.0	48.3	50.7
1890	120.0	146.0	59.0	86.6
1900	147.0	158.0	65.0	132.9
1910	187.0	157.0	72.8	184.0

Sources: as Table 4.1

teaching methods was felt. Other small European countries (Switzerland, Netherlands, Sweden) had a great number of public schools and teachers by the mid-nineteenth century. In Great Britain, on the other hand, the conditions of the industrial revolution meant that 'popular education remained stagnant (in the first decades of the nineteenth century), whilst the economy expanded and wealth mounted now that the proportion of disposable income destined to education diminished progressively. Accumulated wealth was more often employed in contracting growing masses of children in factories than in sending them to schools' (Cipolla, 1969). In Britain, education was still in the hands of municipalities and parishes, or philanthropic societies that created the 'mutual education' of Bell and Lancaster. Only from the fourth or fifth decade was Great Britain conscious of the necessity to systematize education with state intervention. The relatively late state intervention in this country, in comparison with other European countries, would have an effect upon the process of institutionalization of some sciences and especially upon geography.

During the nineteenth century, the impact of the work of Enrico Pestalozzi (1746–1827) contributed to the spread throughout Europe of a concern for the change of pedagogical methods and for an active education, not a pedantic one, in which the child would substitute his own personal experience for books. Form (drawing and geometry), number (arithmetic) and name (language, vocabulary) were for Pestalozzi the basic elements of the effective 'intuition' of things: the study of the sciences, geometry, mathematics, drawing and language would be basic in the Pestalozzi centres. Along with these, contact with nature, clearly of

Rousseauian and Physiocratic roots, became another of the basic principles of the new pedagogy, also prompted – accentuating the idea of the profound unity of the real – by the work of Friedrich Fröbel, the great pedagogue of German romanticism (Abbagnano and Visalberghi, 1957, 466–87). Both authors are found at the root of the development of the *Heimatkunde*, which had such an effect upon the teaching of geography.

Linked with the above currents, from the middle of the century, was the influence of positivism in pedagogy which led to an insistence on attentive and scrupulous observation, on experiment, and on the generalization of methods that proceeded from the particular to the general.

Geography, one of the sciences that traditionally formed part of the teaching programmes, was affected by this pedagogic innovation; through the practice of concerned teachers it became active education, centred upon observation. And, as a science which taught basic notions about our planet, it was particularly to incorporate these new ideas and apply them to the study of the home territory of the child, as a first stage in the study of the earth as a whole.

Geography was thus reinforcing its own position in the educational realm, and becoming a worthy subject for teaching programmes when state intervention and administrative centralization began to formalize these programmes, converting them into legal texts in the greater part of the European countries. The presence of geography in these programmes was also supported in secondary education by the traditional union of geography and history, and by the historians' interest in giving knowledge of the 'theatre' in which developed the historic acts.

Geography and nationalism

Despite enjoying the support of tradition and despite the innovation in teaching methods, geography had clear rivals that aspired to its function. The development in Great Britain around 1870 of the *Physiography* of Thomas Huxley, a naturalist who had felt the impact of the Darwinian revolution (Stoddart, 1975a), is a good example of this threat.

It is reasonable to imagine that during the nineteenth century, geography could have been displaced in primary and secondary education and been substituted by various social and natural sciences. Contrary to what one is accustomed to think, the permanence of geography in teaching programmes is by no means inevitable, and as we have seen, it was, in general, a science little esteemed by scientists during the first half of the nineteenth century and even later. The reasons why geography continued to figure in the programmes, despite the direct rivalry of other sciences which were not able to introduce themselves into basic education (one thinks of physiography or ecology, and in the social

sciences economics or later sociology, a science of a strong development
and high prestige), needs explanation.

Among the reasons that explain the triumph of geography over rival
disciplines, there is one of great importance: the function assigned to
geography in the shaping of a feeling of nationalism.

The development of European nationalism during the nineteenth
century is related to innumerable factors (the influence of the French
Revolution, romanticism, discontent among small agricultural land-
owners and the emerging middle class, etc.). It is evident, nevertheless,
that to a great extent nationalism also coincides, above all after 1848, with
the interests of the growing European bourgeoisie, at a time when
national markets were united and the bourgeoisie began dismantling the
structures of the old régime and of rationalization and homogenization of
their respective territories of influence.

In the development of the feeling of nationality, the idea of 'patria', or
knowledge of the history and geography of one's own country, was an
indispensible element. 'One only loves what one knows' would be an
idea shared by politicians and pedagogues. This explains why both
sciences could enter the basic education programmes with the strong
support of the state, in that their presence corresponded with political
interests.

Geography fulfilled a role which – like that of history – was absolutely
essential in the epoch of the appearance of the European nationalism.
And it fulfilled it not only through teaching, transmitting to the populace
ideas of the 'unity within diversity' of the national territory, but also by
way of 'scientific' and popular works about their countries, works whose
suggestive titles (like *La Patria*) left little doubt about their objectives.
When, in the second half of the century, geographers began to consider
their science as a discipline concerned with the relations between man
and environment, the magnification of the ties and dependencies that
united man and territory would only reinforce the role of geography.

Thus, geography, during the second half of the nineteenth century,
increasingly developed into a science at the service of governmental
interests, of the nationalist European bourgeoisie, who gave generous
support to successive ministries of public education. Despite this support, the official geography (there also existed a marginal geography)
created a new group of adversaries: those who advocated the suppression
of frontiers and international solidarity. The following passage of the
geographer Marcel Dubois, delivered at the Sorbonne in the presence of
eminent French officials during the inaugural lecture of the colonial
geography course in 1893, is sufficiently clear in this respect.

I know that there exists a group, happily small, of historians that
with difficulty pardon geography the fact that it marks a link between

man and the soil, because this same idea of localization, of mutual influence, is something similar to the idea of 'patria', upon which they have declared war. Doubtlessly, geography molests their propaganda; because the chimera of the suppression of frontiers and homelands collides with the reality of the causes that keep groups of men separate. Because in reality, geography has the bad, although I would prefer to say the good, fortune to obstruct the path of these declared or dissimulated enemies of the idea of 'patria'. They have sworn to demonstrate that a certain sociology could completely substitute the role of geography; because they need, for their combinations that have nothing to do with science, an abstract man, always the same, removed from the action of the complex influences of nature; and I like to hear, from the mouths of those that ordinarily do not reproach themselves for their excesses of religious or moral orthodoxy, that geography is accused of being no more than a school of materialism and fatalism. (Dubois, 1893, 129)

One can rarely read such a clear description of the role played by the official geographical science and of the nature of the opposition and polemic that on occasion confronted geographers with social scientists of a left ideology and internationalist ideas (which frequently, although not always, went hand in hand).

To this it must be added that in the epoch of European imperialism, geography fulfilled another important function, spreading knowledge of the colonial empires. The detailed study of the colonies was an indispensable and fairly extensive part of the courses dedicated to 'universal descriptive geography'. The reasons for this outstanding presence are clear, if one takes into account the words of Emile Levasseur – the principal instigator of educational reform in France after 1870 – in the paper which he presented to the International Geographical Conference of London in 1895. To justify his ideas on the teaching of geography at the primary and secondary levels, Levasseur insisted that:

It is very important that the student has a precise knowledge, if not detailed, of the colonies, that it is necessary that he considers them an integral part of his country; the more familiar he is with this idea, the more willing he will be to serve in them or inhabit them without feeling an expatriate. (Levasseur, 1895; cited by Torres Campos, 1896, 217)

Once more, the submission of geography to the interests of the dominant classes in the Europe of the end of the century is manifest, and in this case, significantly, from the mouth of one of the main authors of the institutionalization of French geography. This substantially explains, we believe, the support of governments for this science, so little valued by other scientists, and its presence in all the programmes of primary and secondary education.

Thus, geography had the privilege to be one of the sciences favoured in the movement for educational reform and expansion. To provide teachers, geography became institutionalized in the university. And thus was born the scientific community of geographers, to teach geography to those who had to teach geography.

The role of the geographical societies

In the process of institutionalization of geography in the university, other secondary factors also had effect. One of these factors stands out – the pressure of the geographical societies for recognition of the academic status of geographical science. But this only reinforces the idea of the close relationship between the institutionalization of geographical science and the interests of the dominant classes, in that the development of the geographical societies is very much linked to the process of European imperial expansion.

The appearance of these societies follows the pattern of the expansionist politics of the European states. As a clear forerunner, one can point to the African Association for Promoting the Discovery of the Interior Parts of Africa, created in London in 1788. The first societies founded were the Société Géographique de Paris (1821) which survived half-heartedly until mid-century, the Gesellschaft für Erdkunde of Berlin (1821) and the Royal Geographical Society of London (1830), which numbered from the start 460 members (Freeman, 1961, 50). Despite these early beginnings, the number of geographical societies grew slowly from the foundation of the first in 1821 to 1865, when there existed only 16 societies. From 1865, however, their growth in number was spectacular. From year to year, new societies were created in a great number of countries: in 1873 alone five new ones appeared; seven in 1876, four the following year and eight in 1878. By then there existed a total of fifty societies, according to a study made in that year and listed in the *Boletín de la Sociedad Geográfica de Madrid* (1879, no. 1). The total membership of these 50 societies was 21,263, the largest being the Royal Geographical Society of London with 3,334 members; most however had a membership of less than 1,000 (table 4.3).

The figures on the evolution of the number of members of some societies demonstrates the existence of a period of crisis, which reached its peak in the decade 1840–50, and which coincides with the situation of general crisis of geography that we have described above; it is from the years 1865–70 that interest in geography really seems to increase. Thus the Geographical Society of Paris, which shortly after its foundation had 378 members (in 1827), experienced later a process of decline that lowered

TABLE 4.3

Classification of geographical societies existing
in 1878, according to membership.

Members in individual societies	Number of societies	Total number of members
More than 2,000	1	3,334
From 1,000 to 2,000	4	5,420
From 500 to 1,000	10	7,002
From 100 to 500	17	4,759
Under 100 members	18	748
Total	50	21,263

Source: Elaborated from the figures of the *Boletín de la Sociedad Geográfica de Madrid*, 1879, No. 1, 273-6.

its membership to 100 in 1850, reaching 200 in 1860; from this date there was uninterrupted growth: 640 members in 1870, 800 in 1872, 1150 in 1875 and 1700 in 1877 (Broc, 1974, 550–51). A similar evolution took place in the Royal Geographical Society of London, which after a brilliant beginning, decayed noticeably until the decade of 1840, coinciding with the British social and economic crisis, when it seemed to be on the point of dissolution. Only after 1851, during the second presidency of R. I. Murchison, it began a trend that increased its membership from 600 to 2,000 towards 1870 (Crone, 1970, 29–30).

In the years following 1878, the founding of geographical societies continued with greater intensity all over the world, coinciding with the period of the apogee of European imperialism. Around 1885, there existed 94 geographical societies (of which 80 were European), with a total of 50,000 members. In 1896, the number of societies had risen to 107; of these, 48 were in France, 42 in Germany and 15 in Great Britain (Freeman, 1961, 53).

Between 1890 and 1920, geographical societies were formed at a slower pace, although in the decade 1920–30, the movement reached a last moment of splendour, with the creation of 30 new societies (see table 4.4).

The curve showing the number of societies existing from 1821 to 1835 has a clear configuration of a graded logistical type (figure 4.1), which is so characteristic of scientific growth in general (Price, 1963; Crane, 1972). A

TABLE 4.4

Pattern of the creation of geographical societies

Period	Number of societies founded
1820–1829	2
1830–1839	4
1840–1849	2
1850–1859	6
1860–1869	6
1870–1879	34
1880–1889	28
1890–1899	10
1900–1909	11
1910–1919	10
1920–1929	30
1930–1940	2
Total	145

Sources: To 1878: Behn in *Geographisches Jahrbuch* (cited in *Boletin de la Sociedad Geográfica de Madrid*). From 1879: Sparn, E. in *Gaea, Boletin de la Sociedad Argentina de Estudios Geográficos* No. 22 (cited by J. Gavira, 1948).

Notes: The data from Sparn refers to the societies existing in 1945, which explains the differences that exist between the data of Behn and Freeman.

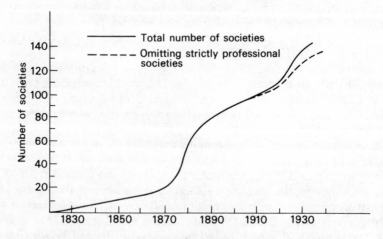

FIGURE 4.1   Evolution of the number of geographical societies 1821-1935

period of slow growth between 1821 and 1870 is succeeded by two decades of exponential growth in which the number of societies is doubled approximately every 10 years. In 1869 there were 20 societies, in 1877 there were 40, and in 1889 their number exceeded 80. There followed a new period of slow growth in which the curve approaches the limit of saturation. This phase is interrupted by a new period of rapid growth, that nevertheless does not reach the same intensity as that of the period 1870–90. To interpret this intensification in the pattern of growth, one must take into account two facts: the societies created from 1910 appear in non-European countries (11 societies of a 'classical type'); and on the other hand, we are now dealing with professional associations (societies of geography teachers, of academics, of scientists specialized in various branches) and which therefore possess a very different character from the geographical societies of the nineteenth-century 'classical' type.

Most of these societies originated in European countries, but from the nineteenth century they also began to appear in the new American nations or in the territories under direct European rule. Towards 1935, the 136 societies in existence were distributed as shown in table 4.5.

European imperialism was without doubt a driving force for this extraordinary growth of geographical societies in the period after 1870. As Lord Aberdare said in 1885:

to the politicians of all the great European nations, the period has been one of intense interest and anxiety, connected more or less

TABLE 4.5

Distribution of geographical societies by continents (1935).

Continents	Number of societies	Number of members
Europe	91	76,182
America	25	17,527
Asia	13	4,026
Africa	5	3,959
Oceania	2	2,018
Total	136	102,712

Source: Sparn, E., cited by Gavira, J., 1948, 300.

with questions of vast territorial requisitions. To the geographer the
interest, although less painful, has hardly been less keen. (Cited by
Freeman, 1961, 58)

The member of these geographical societies were professional soldiers,
naturalists, navigators, merchants, politicians, clergymen, etc. For
example, among the founders of the Royal Geographical Society of
London were: the Secretary of the Admiralty, a navy official specializing in
hydrography, the director of the Ordnance Survey, the president of the
Royal Astronomical Society, a diplomat, a botanist, a linguist, an
antiquarian and a geologist (Crone, 1970, 27). Only at the end of the
nineteenth century, in other words when geography had become
institutionalized, did the number of professors and academics begin to be
significant. Relations between the scientific community of geographers,
once established, and the geographical societies, were, however,
generally less than cordial.

The objectives of these societies were much wider than what today is
understood as geographical, and included not only the organization of
exploration and the development of commerce, but also the creation of
meteorological stations, the making of astronomical observations, ethno-
graphic studies, etc. To achieve their goals, some societies had access to
large budgets from their membership fees and many enjoyed direct state
economic support. But state support could also take other forms: aid for
their publications, subscriptions to journals by official organizations,
financing for particular projects, etc. The active participation of some
politicians (ministers, senators, etc.) in the societies assured a valuable
channel of communication for getting this support. And the interest of
the politician in the work of the societies was doubtlessly great because,
as the motto which headed one geographical magazine declared, 'The
Earth will belong to he that knows it best.'

The societies acted through the organization of conferences, the
concession of prizes and medals (large medals for explorers, and prizes
for books or scholars that stood out), the organization of exploration or
the granting of aid to make them possible, defending and supporting the
idea of colonial expansion (applauding all the campaigns of expansion,
for example). They were also the place where explorers were obliged to
publicize the results of an expedition. The activity of the societies was
shown in their periodical publications (magazines or bulletins) and in the
series of scientific publications that they edited or patronized.

The idea of contributing to the expansion of civilization (European, of
course) was accepted by all and was an ideological justification of the
exploration ventures that preceded those of conquest and civilization. As
Freeman correctly writes,

the need to spread civilizing influence over the more barbarous parts of the world was widely accepted by public opinion at this time. Geographical societies not only satisfied a natural curiosity about the more savage aspects of nature and society, but also cast a shrewd glance at the eventual possibilities of trade and colonial expansion. (Freeman, 1961, 51)

Thus, it is from these societies that geographical science received strong support in its institutionalization process. In a good number of European countries, the geographical societies actively pressurized governments for the university institutionalization of the science. This happened, for example, in France, where the national congresses of geographical societies repeatedly voted for this purpose (Broc, 1974). Similarly, in Great Britain, after the report by John Scott Keltie on the state of geographical teaching in Europe, the Royal Geographical Society offered aid to the universities of Oxford and Cambridge for the creation of posts for geography professors (Scargill, 1976; Stoddart, 1975b). And the Italian Geographical Society likewise played an equally decisive role, as we shall later see, as did the Geographical Society of Madrid, which from 1880 made repeated reports to the Ministry of Public Instruction on the convenience of creating university professors of geography.

The growth of the community of geographers

Because of the factors indicated, during the second half of the nineteenth century the institutionalization of geography proceeded and the scientific community of geographers appeared. Here, not in Humboldt or Ritter, is to be found the origin of modern geography; from this period it became possible to produce and reproduce scientific knowledge at a social scale, which does *not* necessarily imply that the science produced possessed an internal coherence.

The creation of university posts allowed this process of institutionalization to continue. From the description above, one can expect the existence of a close relationship between the growth of basic schooling and the corresponding demand for geography teachers and the creation of university chairs. Although this is a theme that requires more investigation, the available data seems to indicate that this correlation exists.

Thus, the early development of university geography in Germany and Switzerland may be interpreted in the light of the fact that these countries were those which first declared elementary education obligatory, and achieved important advances in popular education from the beginning of the nineteenth century. In France, the defeat of 1870 by the Germans revealed not only the potency of German industry and army, but also the power of German science and culture. The idea then spread that 'it was

c

the German teacher who won the war' (Meynier, 1969, 8) and this contributed to the start of profound educational reform that is found at the root of the development of French geography. These three countries were the first foci of the development of modern geography and in 1875 there existed chairs in this science in seven German universities, seven French and three Swiss, according to a report of the Royal Geographical Society (Stoddart, 1975b).

The decade that followed 1875 saw the appearance of a great number of new university chairs. Another British report, that of John Scott Keltie in 1886, indicates that the number of chairs existing on the continent rose to 45. At that time there still did not exist a chair of geography in the British universities, which perhaps can be seen as a consequence of the delay in the process of public education. In any case, the slow university institutionalization of geography in Great Britain invalidates the interpretation that relates the development of modern geography to colonial exploratory activity: if any country had colonial interests it was Great Britain in the nineteenth century, which makes it difficult to understand, from this point of view, the delay of British geography

During the 10 or 15 years following, university institutionalization of geography took place in Great Britain and the United States. In the latter, despite the early foundation of the American Geographical Society (1851), the real foundation of modern geography took place in the decade of 1890, coinciding with a wide movement of reform of secondary education (Dickinson, 1976, 184).

The decade of the 1890s may be considered that of the coming of age of the scientific community of geographers, reflected in the International Geographical Congresses that they held. That of Bern (1891) was the congress in which the geography professors, the real members of the scientific community of geographers, are found for the first time in the majority at an international congress of geography.

The participants of the first congresses of geography belonged in fact to various socio-professional groups: travellers and explorers, cartographers, meteorologists, professors, journalists, soldiers, and historians. At first, geography interested very definite social groups (the military, politicians, explorers) and scholars who had not yet formed their own institutionalized scientific communities. A statistical analysis made by J. J. Dubois (Union Géographique International, 1972) demonstrates the variety of the socio-professional composition of the participants (table 4.6).

The first congresses were characterized by the presence of relatively few geography teachers (secondary and higher education), which was to be expected in those years of relatively weak institutionalization of the science. Politicians and the military constituted the largest proportion in the congress, together with merchants and industrialists and students of

TABLE 4.6

Socio-professional categories of participants in the International Congresses of Geography (per cent).

	Professors	Geographical Society members	Explorers	Related sciences	Engineers	The military, diplomats, ministers	Merchants, financiers, industrialists	Various	Total
I Amberes, 1871	22.1	—	4.2	18.2	8.3	30.8	16.4	—	100
II Paris, 1875	18.2	—	3.2	13.5	13.8	34.5	17.1	—	100
III Venice, 1881	32.5	28.8	—	5.6	3.2	15.6	—	14.3	100
V Bern, 1891	57.4	—	—	5.7	5.7	19.3	—	51.9	100
XX London, 1964	85.3	—	—	7.9	0.6	4.4	—	1.8	100

Source: Dubois, J. in Union Géographique International, 1972, 50–4.

Note: The percentages have been calculated from the number of participants whose professions are known. For some categories information is not available for all the congresses.

related sciences (cartographers, topographers, meteorologists, hydrologists, statisticians, historians, and economists; the proportion of the last group diminished, however, through their integration into other institutionalized scientific communities or through their conversion into professionals of the geographical science (geologists who became geographers, etc.). Under the heading 'Various' (table 4.6) are included: liberal professions, journalists, museum curators, archivists, librarians; among them were also found editors (in the Congress of Bern, 1891, 41 persons out of a total of 293 whose professions are known), which demonstrates the interest they had in a science that always has had widespread popular demand (travel books, encyclopaedias, etc.), and the growing market for textbooks through geography's presence in the curricula of primary and secondary schools.

The proportion of geography teachers was increasing and became a majority in 1891. The tendency continued, with some oscillations, in the later congresses in which the proportion reached 85 per cent in the XX Congress held in London (1964). This confirms, from another perspective, the almost exclusively academic character of the scientific community of geographers.

Strategies of the community of geographers: towards a model

The institutionalization of geography in the university posed problems of recruitment of new professionals, since geographers did not at that time exist. For this reason, the university professors of the newly institutionalized science – those who became the founders and, in time, 'leaders' of the new community – came from diverse disciplines: historians, geologists, ethnographers, zoologists, naturalists. Perhaps some suddenly realized that it was worth devoting their energies to this new and attractive science. Others, undoubtedly numerous, came to geography after a careful evaluation of the professional opportunities that it offered. The study of the motivations that influenced these 'conversions' to geography and the entry into the scientific community of geographers constitutes an urgent task, although it is not hazardous to state that, in many cases, professional opportunity was the decisive element – similar to what also occurred in the development of other disciplines (see, for example, Ben-David and Collins, 1966). Faced with the traditional hagiographic interpretation that customarily shows the conversions as the fallen on the trip to Damascus, as sudden revelations of a new and marvellous science – a vision revealed at times during a voyage undertaken with other ends – the impartial reading of personal testimonies permits a different version. And this not only applies to second-rate geographers, but also includes the dominant figures. Sufficient, as an example, are the words of Ratzel, recorded by Jean Brunhes.

Professor Ratzel himself told me, in January 1904, a few months before his death, the characteristic evolution of his career, in these terms: 'I have travelled, drawn, described, which led me to the *Naturschilderung*. Meanwhile, I returned from America and *they told me that they needed geographers*. Thus, I collected and coordinated all the facts I had observed and collected on the Chinese immigraion to California, Mexico and Cuba and wrote my work on Chinese immigration, which was my practical thesis. (Brunhes, 1912, 43; italics added)

The case of Ratzel is very similar to that of Richthofen, Vidal de la Blache and so many other geographers who in the last three decades of the nineteenth century occupied university chairs of a geographical science which offered clear professional opportunities.

Once within the community, the newly arrived doubtless experienced a crisis of identity, and they were concerned, in some cases, to mark clear distances with respect to their former colleagues, emphasizing that which was distinctive in geography with respect to the science which they had left. This explains why, surprisingly, some of the new geography professors tended to work on aspects of geography totally different from those of their prior education: ethnographers or historians became physical geographers, geologists developed the human aspects of the new science, etc. Others adopted a different strategy, tending to assert the knowledge of their initial training to create a place of prestige for themselves in the heart of the community of geographers; this would, however, need careful balancing, as it was always necessary to justify the new science and emphasize the differences between it and their old one.

In any case, the scientific community of geographers all began working hard to justify their science, trying on the one hand to emphasize its excellence and utility and on the other to delimit it unequivocally from neighbouring disciplines, with whom it had maintained close relations in the past (like history or geology) or with respect to other emergent sciences whose object was partially similar (for example, physiography or ecology). In this task of self-affirmation, geographers saw themselves forced to insist on the intermediate position of geography, on its character of a bridge discipline between the natural and social sciences.

Unlike historians and social scientists, geographers insisted on the 'natural' character of their discipline, and physical geography became the 'soul of geography', following an expression used by French geographers at various times in their disputes with the historians. Unlike the naturalists on the other hand, geography was presented as a science that also must include human data in its descriptions, and later, as an historical and social science. This led to the defence of a conception of geography in which it was given a 'real and proper coordinating function

for scientific notions', as an Italian geographer said in 1895, while
defending the indispensable role of geography in elementary education,
alleging that 'knowledge that could not enter by the door (through
specialized sciences absent at this level of education) will enter through
the window, by way of geography' (see Capel, 1977, 14). The problems
which geographers considered had to be the object of theoretical discussion, and generate scientific output which would show – with more or
less success – the coherence of the new science. This scientific output could
be interpreted as a justifying and rationalizing theoretical development,
achieved from an acquired socio-professional position, which it had no
intention of renouncing.

Naturally, this theoretical output developed using scientific and
philosophical ideas found in the cultural medium in which it occurred. At
the time of the institutionalization of geography, these ideas were
positivism, as a scientific method, and Darwinian evolutionism.
Geography therefore shaped itself as a positive science. In that period,
traditions were also created (accepted ideas about the object of geography
and methods of work), that made it conserve this character for several
decades, with the modifications and adaptions that historicism imposed
on the discipline at the end of the century.

From the beginning of the twentieth century, the community of
geographers appears, generally, consolidated and recognized within the
scientific communtiy. The principal task was to defend geography from
other developing rival sciences. It is the period in which the polemics of
geographers with other scientific communities that try to deprecate
geography (like the sociologists or ecologists), acquired great virulence.
The struggle with these rival communities is not a simple theoretical
question. It is also – perhaps above all – a struggle to dominate other
professional fields. When the encounters did not lead to the annihilation
of one of the communities, one arrived at a situation of respect for
respective professional fields of interest. This normally generates continuous scientific output tending to carefully delimit the respective
sciences. That a diploma qualifies one for such and such profession can be
at the base of hairsplitting discussions on the subtle differences between
neighbouring sciences, differences which are, in truth, often nonexistent
to a reasonable observer, who is not affected by the socio-professional
implications of the discussion.

The invasion of some of the socio-professional fields of influence by
rival communities generally provokes violent opposition in the affected
community, translated into scientific articles dedicated, in one way or
another, to this theme. Two types of reactions take place in the violated
community: defensive reactions, which try to deprecate the rivals and
show that they are not capable of performing the pretended function; and
offensive reactions, that tend to find new professional outlets, although

this may provoke new confrontations with another community. The polemic of the years 1920–30 sustained between geographers and specialists in the natural sciences as to who should teach it, is an example of the first kind of reaction; while the inclination towards land use and 'applied geography' after 1930 can be interpreted as an offensive reaction, which would lead geographical science through new paths.

This constant battle with neighbouring disciplines generates concern that the science should not be revealed as backward and deprecated in front of others. This would lead to the incorporation of ideas, methods or theories from other sciences, without the scientific community of geographers seeming to worry about the fact that these new ideas were not coherent with conceptions previously held.

Finally, one must take into account that the same scientific community is subject to violent tensions, as a result of the implacable struggle for prestige and – ultimately – for power. The confrontations between different scientific ideas or between distinct scientific paradigms can be the occasion in which these oppositions come to light and give place to defensive strategies or strategies of rejection – as Taylor (1976) has detected regarding the quantitative debate in British geography.

One can identify these strategies (strategies of self-affirmation, of delimitatión of fields; of searches for new professional outlets of assimilation of the advances of other sciences; and of the struggle for power) from a re-reading of the theoretical output and the words of the members of the scientific community of geographers. The analysis of various national schools permits the checking of the validity of the model presented.[2]

CONCLUSION

The scientific community of geographers is an example of a scientific community constituted from clearly social factors, and not as a result of specific necessities in scientific knowledge. After a period in which geography had entered into profound crisis, suffering a process of depreciation related to the appearance of more specialized branches, the presence of this science in programmes of primary and secondary education generated, from the mid-nineteenth century, a need for geography teachers, which provoked in turn the university institutionalization of the science.

This is the origin of the scientific community of geographers, which may be considered fully constituted and recognized in many countries from the end of the nineteenth century. Institutionalization generated a process of development of geographical science similar to that of other disciplines. As Sklair (1973) has pointed out, 'when a social activity, like a science, institutionalizes itself, it experiences very significant changes in its scale, internal relations and in relations it possesses with other spheres or important institutions in the society.'

A large proportion of the members of the new community came to it after a clear appreciation of the socio-professional opportunities, and not for any 'inner call' or vocation towards the science. Although later, certain 'heroic myths' (Taylor, 1976), like that of vocation and sacrifice for the science, can be spread and accepted by the whole community.

Every scientific community, once institutionalized, establishes its own norms and value systems, which become a cohesive and stabilizing element of it. Increasingly, before becoming part of an institutionalized community (here geography), a period of training, of disciplining, is demanded, for the inculcation of those norms and values that were close to the science's heart, of knowledge of the relevant problems and methods of approaching them, of precedents and the work of the founding fathers (von Humboldt and Ritter).

The established community employs strategies tending to reproduce and amplify itself. Never will it opt for self-liquidation: the community will defend its survival, even if other communities of scientists investigate similar problems with like method, or if the logical incoherence of the conceptions that they defend is revealed. One can express the hypothesis that entire communities have rested, at times, upon hardly coherent conceptions, sometimes incoherent, without the community's members having thought for a moment to question their existence. The case of the scientific community of geographers (like that of the anthropologists) could, perhaps, be formulated from this perspective. Everything will be sacrificed for the reproduction and growth of the community, including the coherence of the very conception of the discipline: different conceptions can defend themselves in distinct moments or even simultaneously, without putting into doubt the continuity of the science practised.

In the heart of the community, hierarchically structured, the struggle for prestige and power creates, on occasions, strong internal tensions and more competition than cooperation. Despite this, the members of the group appear united by strong ties of solidarity facing rival communities.

It seems clear that at least some scientific communities, like that of the geographers, have been strongly influenced by social factors. Despite all, communities segregate their own traditions and solidarities and become social structures of a certain autonomy, which can react rigidly before external pressure or defend themselves through adaptive processes.

The thesis of this paper is that the strategies of the scientific communities can be detected through the 'scientific' output of its members. But also that *the same scientific output – as much theoretical as of specific research – can be likewise interpreted as results of the same strategies and not only as the logical and inevitable product of the development of scientific knowledge.* The 'internalist' conception of science, that considers it as an autonomous intellectual result of purely cognitive processes – the concept that appears in the works of Weber, of Popper, of Koyré and, in a lesser way, of Kuhn – must be modified. The evolution of 'scientific knowledge' is not only the result of the contrast and verification of theories and hypotheses, of the rigorous critique of scientific conceptions, that produce the devaluation of some and the triumph of others, of the battle between alternative paradigms. It is also the result of socio-professional interests of the members of the scientific community and of the strategies that they use to defend their interests against those of rival communities, and also of the struggle for power in the heart of the community. The confrontations of rival, alternative paradigms and the choice between them can be not only a battle between scientific conceptions that, for better or for worse, resolve determinate problems, but also a time of confrontations in the very heart of the community. The reasons why a scientist adheres to or rejects certain theories or conceptions is a question which one probably must answer by alluding not only to scientific reasons, but also socio-professional motivation. The production and the evaluation of scientific knowledge also relate – along with other factors – to the social structure of the science.

In the debate on the theory and sociology of science, the accent is frequently placed on the triumph of new and innovative ideas. But the persistence of old and unsustainable conceptions also deserves attention. As much in the triumph of new ideas as in the persistence of the old, factors of a social character seem to have an important influence. For this, the analysis of the social structure and of the strategies of the scientific communities and the relation of scientific output with these strategies, is a theme of great significance for the future, in that it can throw light on the ways in which scientific knowledge is produced.

Notes

[1] This paper was translated into English by Kirk Mattson.
[2] We omit here the analysis of the case of Italy, which may be found in Capel, 1977: *Geo Critica*, 9.

References

Abbagnano, N. and Visalberghi, A. 1957: *Linee di storia della pedagogia* (Turin).
Almagìa, R. 1961: *La geografia in Italia dal 1860 al 1960* ('L'Universo', Instituto Geográfico Militar, Florence), 419–32.
Beck, H 1959–61: *Alexander von Humboldt*. 2 vols. (Wiesbaden).
Ben David, J. and Collins, R. 1966: Social factors in the origins of a new science: the case of psychology. *American Sociological Review* 31, no. 4, 451–65.
Bertacchi, C. 1892: Delle vicende e degli ordinamenti dell' insegnamento geografico nelle scuole secondarie, dalla costituzione del Regno; e proposte dei mezzi per

migliorarlo. In *Atti del Primo Congresso Geografico Italiano* (Genoa), vol. 2, part 2, 551–83.

Broc, N. 1974: L'établissement de la géographie en France: diffusion, institutions, projets (1870–1890). *Annales de Géographie* 83, 545–68.

Broc, N. 1976: La pensée géographique en France au XIX siècle: continuité ou rupture? *Revue Géographique des Pyrénées et du Sud-Ouest*, 47, 225–47.

Brunhes, J. 1912: *La géographie humaine, essai de classification positive, principes et exemples* (Paris).

Capel, H. 1974: La personalidad geográfica de Varenio. In Varenio, B., *Geografia general en la que se explican las propriedades generales de la Tierra* (Barcelona), 11–84.

Capel, H. 1977: Institucionalización de la geografía y strategias de la comunidad científica de los geografos. *Geo Critica* (Barcelona) 8 and 9.

Cipolla, C. 1969: *Literacy and development in the West* (London).

Claval, P. 1964: *Essai sur l'évolution de la géographie humaine* (Paris).

Crane, D. 1972: *Invisible colleges: diffusion of knowledge in scientific communities* (Chicago).

Crone, G. R. 1970: *Modern geographers. An outline of progress since AD 1800* (Royal Geographical Society, London).

Dickinson, R. E. 1969: *The makers of modern geography* (London).

Dickinson, R. E. 1976: *Regional concept. The Anglo-American leaders* (London).

Dubois, M. 1894: Méthode de la Géographie coloniale. Leçon d'ouverture du cours de Géographie coloniale, Faculté des Lettres (1893). *Annales de Géographie* 3, 121–37.

Freeman, T. W. 1961: *A hundred years of geography* (London).

Gavira, J. 1948: Las sociedades geográficas. *Estudios Geográficos*, 309–15.

Gerasimov, I. editor, 1976: *A short history of geographical science in the Soviet Union* (Moscow).

Hall, D. H. 1976: *History of the earth sciences during the Scientific and Industrial Revolutions, with special emphasis on the physical geosciences* (Amsterdam).

Hartshorne, R. 1959: *Perspectives on the nature of geography* (Chicago).

Hettner, A. 1898: Die Entwicklung der Geographie im 19 Jahnrhundert. *Geographische Zeitschrift* 4, 305–320.

Kretshmer, K. 1930: *Historia de la Geografía* (Barcelona).

Matveyeva, T. P. 1976: Documentary sources for the study of the development of geographical thought. In *History of Geographical Thought* (Abstracts of Papers, Leningrad, International Geographical Union), 55–9.

Meynier, A. 1969: *Histoire de la pensée géographique en France (182–1969)* (Paris).

Minguet, C. 1969: *Alexandre de Humboldt, Historien et géographe de l'Amérique espagnole (1799–1804)* (Paris).

Pracchi, R. 1964: Studi generali sull'Italia e monografie regionali. In *Un sessantenio di ricerca geografica italiana, Memorie della Società Geografica Italiana*, XVI, 575–600.

Price, D. J. de Solla 1963: *Little science, big science* (NewYork).

Sanderson, M. 1974: Mary Somerville: her work in physical geography *Geographical Review* 64, 410–20.

Scargill, D. I. 1976: The R.G.S. and the foundations of geography at Oxford. *Geographical Journal* 142, 438–61.

Schaefer, F. K. 1953: Exceptionalism in geography: a methodological examination. *Annals of the Association of American Geographers* 43, 226–249.

Sklair, L. 1973: *Organized knowledge* (London).

Stoddart, D. R. 1975a: 'That Victorian Science': Huxley's *Physiography* and its impact on geography. *Transactions of the Institute of British Geographers* 66, 17–40.

Stoddart, D. R. 1975b: The R.G.S. and the foundations of geography at Cambridge. *Geographical Journal* 141, 216–39.

Sukhova, N. G. 1976: The idea of interaction between natural phenomena on Earth's surface and the development of geography in Russia. In *History of Geographical Thought* (Abstracts of Papers, Leningrad, International Geographical Union) 62–5.

Taylor, P. J. 1976: The quantification debate in British geography. *Transactions of the Institute of British Geographers* new series 1, 129–42.

Torres Campos, R. 1896: *La geografía en 1895. Memoria sobre el VI Congreso Internacional de Ciencias Geográficas celebrado en Londres* (Madrid).

Union Géographique International (1972): *La géographie à travers un siècle des Congrès Internationaux* (Caen).

Valskaya, B. A. 1976: The development of geographical thought in Russia in the first half of the 19th century. *History of Geographical Thought* (Abstracts of Papers, Leningrad, International Geographical Union), 46–51.

The Paradigm Concept and the History of Geography

D. R. Stoddart

It is almost 20 years since T. S. Kuhn introduced the notion of the paradigm in his analysis of *The structure of scientific revolutions* (1962). Since then both term and concept have been widely adopted in philosophical and historiographical discussion, especially in the social sciences, and after some hesitation the paradigm idea is suddenly common currency in geographical writings too. In this essay I wish to review the ways in which the term has been applied, its value in geographical historiography, and especially the reasons why it has had such sudden and widespread popularity.

Kuhn used the term paradigm to denote a generally accepted set of assumptions and procedures which served to define both subjects and methods of scientific enquiry. For him, 'normal science' was carried out within the context of a prevailing paradigm, which itself both defined the importance of questions for study and set criteria for the acceptability of solutions. Within this framework, much of 'normal science' was of a 'puzzle solving' kind, seeking laws, constants, coefficients and other relationships within the context of the paradigm. Such work, while necessarily restricted in its scope, was nevertheless highly focused and usually productive.

But from time to time the goals and procedures within which scientists operated came to appear less satisfactory, and the prevailing paradigm would be replaced by a new. This process of paradigm change thus supplied a key to the interpretation of historical development in the sciences: change came to be seen as episodic, or indeed 'revolutionary'. With a change in paradigm, old problems lost their significance, old methods their relevance, and the focus of research moved abruptly to new areas. 'To desert the paradigm is to cease practising the science it defines' (Kuhn, 1962, 34).

Kuhn's concept, used as a key to the understanding both of the formal structure of investigation and of the interpretation of change in the history of science, was first explicitly introduced into geography by Haggett and Chorley (1967). They defined paradigms both operationally as 'stable patterns of scientific activity' and more formally as 'large-scale models'; and Kuhn's views were then used somewhat confusingly to argue for a 'model-based paradigm' for geography itself. This polemical use of Kuhn's ideas as a means of promoting particular views in geography was accepted implicitly by many commentators at the time (e.g. Saey, 1968), and subsequently Harvey in his influential *Explanation in Geography* (1969, 16-18) used the paradigm idea as an organizing framework. Thereafter the term passed into common use, and has recently been used to structure a recent history of Anglo-American human geography since 1945 (Johnston, 1979) (though with a postscript disavowal).

GEOGRAPHICAL PARADIGMS?

The only existing geographical paradigm discussed by Haggett and Chorley (1967) was a classificatory or regional one, exemplified by Berry's (1964) proposed data matrix and Grigg's (1965) logic of regional systems. Against this they set their own proposal for a model-based paradigm. In their discussion, this latter is clearly an inclusive category, defined in terms of characteristics rather than of content. A new paradigm, they suggest, 'must be able to solve at least some of the problems that have brought the old to crisis point'; it must be 'elegant, appropriate and simple'; and it must contain 'potential for expansion' (1967, 37-8). The samples they cite (derived from work on migration, point patterns, search theory, network analysis and diffusion studies) scarcely compare in scale with Kuhn's own examples of paradigm change (which hinge on the work of Copernicus, Newton and Einstein).

Haggett (1965, 10-13) had, of course, already suggested the existence of what he then termed simply 'widely held views' about the nature of geography: as the study of areal differentiation of the earth's surface, as the study of landscape, as the study of the relationship between earth and man, and as the study of distributions or location; and it is perhaps on this scale, rather than that of specific techniques or theories, that the application of the paradigm idea should be sought. Perhaps in Haggett's third category (though he himself treats it in his second) the work of the Berkeley school under the powerful and inspiring influence of Carl Sauer must come closest to the main criteria set by Kuhn for paradigm recognition. The extraordinary dominance exercised by Sauer, the loyalty he commanded, and the affection and regard in which he was held, are made very clear by Leighly (1979) and Parsons (1979). But the Berkeley school co-existed with other very different research traditions, not only in Europe but also in the United States; and its distinctiveness did not prevent Sauer from twice serving as president of the Association of American Geographers.

Apart from the Berkeley School, the paradigm idea has been used in geography in three main areas. The first and most popular is that of continental drift and plate tectonics (Frankel, 1978; Kitts, 1974; Hallam, 1973; Moffatt, 1977; Vine, 1977). Here commentators have been fascinated by the way in which Alfred Wegener's basically correct conclusions were powerfully opposed for years by physicists such as Harold Jeffreys, because of the difficulty raised over a suitable mechanism for drift, until the discovery of the significance of magnetic lineations on the ocean floor by Vine and Matthews demonstrated the reality of major continental movements. Second, Davisian geomorphology is often seen as establishing a paradigm for the historical analysis of landforms (denudation chronology) which dominated geomorphic work from the end of the nineteenth century until after Davis's death in 1935 and its final overturning by a new concern with process and system studies introduced by the work of Horton (see Chorley, 1965). Third, Haggett's own book on *Locational analysis in human geography* (1965) is often taken to mark a convenient break between an older classificatory and descriptive mode of study and a newer analytical and quantitative one.

Recently, however, the paradigm idea has been used on a variety of scales. Berry (1973) proposed a new paradigm for geography involving action and change, and he has also (1978) discussed the history of environmental determinism in paradigm terms. Meyer (1973) and Herbert and Johnston (1978) apply the idea to urban locational analysis. Garrison (1979) refers in the same paper both to 'causal paradigms' and to 'paradigms for the study of urban areas, transportation, and regional science'. The list could be expanded, and also readily duplicated for neighbouring fields such as economics and sociology. In each case the paradigm terminology has been used to illuminate either the establishment of views of which a commentator approved, or to advocate the rejection of those he did not.

APPLICABILITY?

This ready acceptance of Kuhn's vocabulary has occurred without any close attention to Kuhn's own statements or to the critical literature on them in the history and philosophy of science. We must first enquire whether the paradigm idea is useful in understanding the processes of change in geography on other than a superficial level.

The confusion implicit in Haggett and Chorley's initial usage and the plasticity with which the concept has subsequently been applied mirrors the multiplicity of meanings attached to the idea at different times by Kuhn: Masterman (1970) identifies 21 discrete definitions in his 1962 book alone. But if we leave this to one side, and accept the view of a paradigm as a consensus of aims and methods which defines, until replaced, the pursuit of normal science, then the closer the analysis the less apposite the concept appears in geography.

Thus, for example, whereas it is true that Davisian geomorphology was codified in textbooks and widely adopted by both researchers and pedagogues during the first half of the twentieth century, Davis's views were received with scepticism by many, even in his own lifetime. In Britain, H. R. Mill and J. S. Marr, in the United States Fenneman and Chamberlin, and in Germany Albrecht Penck and Alfred Hettner all declined to be persuaded. Alternative geomorphic systems, notably that of Walther Penck, were not only actively used in research (for example in Sauer's paper on the tectonic landforms of Chiricahua (1930)) but also given textbook authority, as in Von Engeln's influential *Geomorphology* (1942). The Berkeley Department of Geography virtually ignored Davisian methods (Leighly, 1979). Kuhn's formulation scarcely makes allowances for coexisting paradigms (especially coexisting over nearly half a century), and if nevertheless Davisian geomorphology is to be described as a paradigm, then the meaning of the term requires revision. Only by analysing the effects of Davis's views on non-scientists (such as the historian Walter Prescott Webb) and educationalists is it possible to maintain the value of the paradigm idea, but then on essentially social rather than scientific grounds.

A similar analysis could be made of the 'widely held views' – not universally held views – listed by Haggett (1965), as well as of some of the other recent candidates for paradigmacy: 'relevant geography' (see Stoddart, 1975), behavioural geography, and the wide range of attitudes discussed by Gregory (1978). None has been uniquely accepted as a geographic paradigm in Kuhn's original sense, all have a long history within the subject, and all have coexisted with many other divergent schools of thought. Johnston (1978, 201) has indeed concluded from his 'failure to match Kuhn's model to recent events in human geography' that 'the model is irrelevant to this social science, and perhaps to social science in general.' In a similar way, Tribe (1978, 15) has concluded that Kuhn's proposals 'are of no use in considering the problems faced by the history of economics.'

If the concept of the paradigm has thus been used so loosely, on a variety of levels, and without specific reference to Kuhn's own criteria, it is clearly not possible to infer from its use the operation in each case of the processes of scientific change which Kuhn described. But since, even when loosely used, the term carries with it connotations of revolutionary change, it is necessary to consider briefly how change has occurred in some of the geographical cases cited as paradigms.

PARADIGM CHANGE

The importance of the paradigm idea, in Kuhn's terms, lies not only in the way it apparently supplies an interpretative framework for historical studies, but in the implications it carries for why and in what manner change occurs. When change is viewed as paradigm replacement or paradigm shift, large and often unrecognized assumptions are made about the processes involved. Simplistic notions of change imply simplistic assumptions about the behaviour of individual scientists which require consideration before and not after the paradigm idea is accepted. As more is understood of changes in geographical thought over the past hundred years, and particularly of the subtle interrelationships of geographers themselves, the less appropriate and useful does the notion of revolutionary change appear.

The main area of controversy has been that between Kuhn and Popper over the criteria for paradigm rejection (Lakatos, 1970; Blaug, 1976). Essentially Popper emphasizes the importance of methodological procedures, especially leading to the rejection of predictions rather than their confirmation, as the only sound criterion in science; whereas Kuhn's argument is concerned more with the changing attitudes and values of groups of paradigm adherents, not necessarily resulting from any demonstration of error in scientific terms. The problem is particularly intractable even in the more 'scientific' kinds of geographical work, such as quantitative human geography, where low-level predictions (called 'forecasts') are rarely used to test theory, and where the relationship between theory and reality often differs in fundamental respects from that in the physical sciences.

Consider the manner of paradigm change. In the Kuhnian view, this is by a revolutionary process, often expressed in military metaphor: old views are attacked and suddenly 'overthrown'. The introduction of quantification into British geography in the 1960s is seen by Taylor (1976) in this way. In analysing the progress of the revolution, Taylor suggests that the day was carried because mathematics was used by the insurgents as a specialist secret language inaccessible to its opponents, and as intellectual camouflage intended to impress, bolstered by frequent impressive references to Newton, Einstein, Planck, Heisenberg and 'the scientific method', to establish intellectual respectability. The revolution was over, at least according to Burton (1963), by the early 1960s: the old paradigm had been vanquished and a new one had taken over.

Two obvious questions arise. First, what of the quantifiers who had worked happily enough under the old regime? How can their continuing activities be reconciled with this simplistic view of change? I refer not simply to people like Spottiswoode (1861) and Cayley (1879), who managed to publish in spite of the admirals and explorers in the journals of the Royal Geographical Society, but to more central figures such as Christaller and to his many predecessors (Müller-Wille, 1978). Those who say that Christaller was ignored – perhaps even suppressed – in the dark days before 1960 should read their Dickinson (1947). To give another example, there is the clearest foreshadowing of the structure of Haggett's *Locational Analysis* in a remarkable paper published by a leader of the old paradigm (James, 1952). Here James considered the problems of pattern, process and scale, and attempted to reconcile them with the prevailing theme of geography as the study of areal differentiation. And it is not difficult to show how James's views developed over time from his first involvement in technical debates in the *Annals of the Association of American Geographers* in the early 1930s (see, for example, James, 1934, 1937, 1948). In these particular cases the revolution, if there was one, had long and respectable antecedents: the precursors themselves were central figures of the old paradigm. These revolutions were thus processes rather than events, involving a shift in emphasis rather than the wholesale replacement of one set of attitudes by another.

Second, if the revolution succeeded as rapidly as Burton (1963) suggests, why do we find its leading practitioners in disarray, lamenting their 'confusion and doubt' (Berry, 1973, 3), the 'growing isolation of those involved in theoretical and quantitative work' (King, 1979, 157), and the lack of effect of the changes on university admissions policies and curricula (Gould, 1979, 149–50). I raise these points simply to indicate that the revolutionary model of change is self-evidently inappropriate to anything as complex as scholarly endeavour.

But more crucially, the revolutionary model says nothing of the ways in which change is effected: why some views appeal to particular individuals and others do not, why some workers in some localities are attractd by, adopt and transmit new ideas. There are very few studies of the processes and contexts of change in geographical thought. Consider the introduction of the 'new geography' in Britain in the 1880s, said to have been brought about by Mackinder's paper 'On the scope and methods of geography' in 1887 (Unstead, 1949). Examination shows that this paper was very far from being the unheralded frontal assault on the entrenched forces of exploration, which won the day by the force of its intellectual argument. Mackinder's argument had been almost wholly anticipated by others, and was indeed common currency in the Royal Geographical Society in the later 1880s. More to the point, however, its content reflects social and economic as well as intellectual tensions not only in geography but also in neighbouring subjects: the 'new geography' was simply part of a general readjustment of roles and subject matter in the earth sciences at a time of wide educational reform (Stoddart, 1980). The more this complexity is understood, the less revolutionary the process seems and the less dominant a figure Mackinder appears (Stoddart, 1976).

Such an analysis has yet to be made for the progress of the quantitative revolution, though Duncan (1974) has outlined its progress in diffusion terms, and Pred (1979) has described the constant contingencies involved in his own intellectual development. What led Pred to Göteborg in 1960–1 (and ultimately to his association with Hägerstrand), and Harvey to Uppsala and his with Olsson in the same year? The processes of change are concerned with matters such as these, which are simply not considered in a revolutionary view. It is true that the contingencies were somewhat reduced in the emergence of quantitative geography with the organization of summer schools, symposia, and new journals in the 1960s, but these are the central devices of 'normal science', not of revolutions.

It follows from these examples that the adoption of Kuhn's terminology, far from clarifying history, actively distorts it, largely by reducing the participants to caricature figures. Some very clearly become heroes. This is demonstrated to the point of absurdity by Bunge's treatment of Schaefer (1968) and Christaller (1977). Schaefer's personal history and political convictions were unknown (as indeed was Schaefer) to almost every reader of the only paper of any consequence he ever published. Whether Christaller was or was not a fascist is in terms of present attitudes to his contribution of no significance.

Supporters of the old paradigm readily become fools, if not knaves. W. M. Davis, for example, came to be readily considered simply as 'an old duffer with a butterfly-catcher's sort of interest in scenery', as Mackin (1963, 136) describes him. Taylor (1976, 138–9) notes that much of the opposition to quantitative geography came from older men, such as L. Dudley Stamp (1966) (who injudiciously compared the quantifiers to communists) and a succession of presidents of the Institute of British Geographers. But a reading of their presidential addresses gives an impression of cautious sympathy rather than outright hostility. Bunge (1968) again especially identifies Hartshorne as an 'enemy on a personal scale' and a 'lifelong protagonist' (presumably meaning antagonist), without any consideration of the magnitude of his scholarly contribution. It is perhaps worth noting that the 'Planck principle' (Planck, 1948, 22) that new ideas do not convince opponents, who simply die and are replaced by younger and more receptive men, while superficially attractive, has been disproved on the only occasion on which it has been formally tested (Hull et al., 1978).

This is not to say that generalization is impossible about the nature and progress of scientific change and the characteristics of those who are

TABLE 5.1

T. H. Huxley's four stages of public opinion (Bibby, 1959, 77).

I (Just after publication)
The Novelty is absurd and subversive of Religion and Morality. The propounder both fool and knave.

II (20 years later)
The Novelty is absolute Truth and will yield a full and satisfactory explanation of things in general - The propounder a man of sublime genius and perfect virtue.

III (40 years later)
The Novelty will not explain things in general after all and therefore is a wretched failure. The propounder a very ordinary person advertised by a clique.

IV (100 years later)
The Novelty a mixture of truth and error. Explains as much as could reasonably be expected. The propounder worthy of all honour in spite of his share of human frailties, as one who has added to the permanent possessions of science.

---

involved in it. Many such generalizations predate Kuhn's formulation and are independent of it. Thus Huxley long ago suggested a general reaction to any new proposal (table 5.1). Beveridge (1950, 109) repeated much the same idea:

> The reception of an original contribution to knowledge may be divided into three phases: during the first it is ridiculed as not true, impossible or useless; during the second, people say there may be something in it but it would never be of any practical use; and in the third and final phase, when the discovery has received general recognition, there are usually people who say that it is not original and has been anticipated by others.

And there are also interesting generalizations about the innovators themselves (Barbour, 1961). Rogers (1962, 194) in one of the founding documents of diffusion theory, quotes Linton's view that innovators in general are 'very frequently misfits in their societies, handicapped by atypical personalities', and Barnett's that 'the disgruntled, the maladjusted, the frustrated, or the incompetent are pre-eminently the acceptors of culture innovation and change' (see also Roe, 1963). I quote these views in illustration of the fact that the common characterization of the innovator as hero is by no means axiomatically true.

CONCLUSION

Why, if this interpretation is valid, is the paradigm idea so popular, not only in human geography (and other social sciences) but in physical geography too? I suggest that a major reason lies in the way in which the concept of revolution bolsters the heroic self-image of those who see themselves as innovators and who use the term paradigm in a polemical manner, coupled with the fact that Kuhn's terminology supplies an apparently 'scientific' justification for the advocacy of change on social rather than strictly scientific grounds.

In its simplest formulation, the paradigm idea suggests the replacement rather than the testing of ideas, and by extension the replacement of practitioners also: in this lies the heart of Popper's criticism of Kuhn's thesis (Lakatos, 1970). In this sense there is room for sociological enquiry into the extent to which the concept has been used in recent years as a slogan in interactions between different age-groups, schools of thought, and centres of learning, rather than a a useful heuristic model of how and in what manner science is structured and change occurs. We might usefully analyse why some geographers choose to identify themselves as paradigm-changers at the present day, and whether, by their actions since 1960, they have so simplified our perceptions of the processes of change that the paradigm idea comes to appear analytically useful. In other words, those who propound the Kuhnian interpretation have done so in ways which tend to make it self-fulfilling. It is this, rather than its value as a framework for studying historical change, that make the paradigm idea of interest to the historian of science: as itself an object of study, rather than a means of understanding the complexities of change.

References

Barbour, B. 1961: Resistance by scientists to scientific discovery. *Science* 134, 596–602.

Berry, B. J. L. 1964: Approaches to regional analysis: a synthesis. *Annals of the Association of American Geographers* 54, 2–11.

Berry, B. J. L. 1973: A paradigm for modern geography. In Chorley, R. J., editor, *Directions in geography* (London) 3–22.

Berry, B. J. L. 1978: Geographical theories of social change. In Berry, B. J. L., editor, *The nature of change in geographical ideas* (Northern Illinois University Press), 18–35.

Beveridge, W. I. B. (1950): *The art of scientific investigation* (London).

Bibby, C. 1959: *T. H. Huxley: scientist, humanist and educator* (London).

Blaug, M. 1976: Kuhn versus Lakatos or paradigms versus research programmes in the history of economics. In Latsis, S. J., editor, *Method and appraisal in economics* (Cambridge) 149–80.

Bunge, W. 1968: Fred K. Schaefer and the science of geography. *Harvard Papers in Theoretical Geography, Special Paper Series*, A, 1–21.

Bunge, W. 1977: Walter Christaller was not a fascist. *Ontario Geographer* 11, 84–6.
Burton, I. 1963: The quantitative revolution and theoretical geography. *Canadian Geographer* 7, 151–62.
Cayley, G. 1879: On the colouring of maps. *Proceedings of the Royal Geographical Society*, new series, 1, 259–61.
Chorley, R. J. 1965: A re-evaluation of the geomorphic system of W. M. Davis. In Chorley, R. J. and Haggett, P., editors, *Frontiers in Geographical Teaching* (London) 21–38.
Dickinson, R. E. 1947: *City region and regionalism: a geographical contribution to human ecology* (London).
Duncan, S. S. 1974: The isolation of scientific discovery: indifference and resistance to a new idea. *Science Studies* 4, 109–34.
Frankel, H. 1978: Arthur Holmes and continental drift. *British Journal For the History of Science* 11, 130–50.
Garrison, W. L. 1979: Playing with ideas. *Annals of the Association of American Geographers* 69, 118–20.
Gould, P. 1979: Geography 1957-1977: the Augean period. *Annals of the Association of American Geographers* 69, 139–51.
Gregory, D. J. 1978: *Ideology, science and human geography* (London).
Grigg, D. 1965: The logic of regional systems. *Annals of the Association of American Geographers* 55, 465–91.
Haggett, P. 1965: *Locational analysis in human geography* (London).
Haggett, P. and Chorley, R. J. 1967: Models, paradigms and the new geography. In Chorley, R. J. and Haggett, P., editors, *Models in Geography* (London) 19–41.
Hallam, A. 1973: *A revolution in the Earth Sciences: from continental drift to plate tectonics* (Oxford).
Harvey, D. 1969: *Explanation in Geography* (London).
Herbert, D. T. and Johnston, R. J. 1978: Geography and the Urban Environment. In Herbert, D. T. and Johnston, R. J., editors, *Geography and the Urban Environment: progress in research and applications* (Chichester) 1, 1–33.
Hull, D. L., Tessner, P. D., and Diamond, A. M. 1978: Planck's Principle. *Science* 202, 717–23.
James, P. E. 1934: The terminology of regional description. *Annals of the Association of American Geographers* 24, 93–107.
James, P. E. 1937: On the treatment of surface features in regional studies. *Annals of the Association of American Geographers* 27, 213–28.
James, P. E. 1948: Formulating objectives of geographic research. *Annals of the Association of American Geographers* 38, 271–6.
James, P. E. 1952: Toward a further understanding of the regional concept. *Annals of the Association of American Geographers* 42, 195–222.
Johnston, R. J. 1978: Paradigms and revolutions or evolution? Observations on human geography since the Second World War. *Progress in Human Geography* 2, 189–206.
Johnston, R. J. 1979: *Geography and Geographers: Anglo-American human geography since 1945* (London).
King, L. J. 1979: The Seventies: disillusionment and disillusion. *Annals of the Association of American Geographers* 69, 155–64.
Kitts, D. B. 1974: Continental drift and scientific revolution. *Bulletin of the Association of American Petroleum Geologists* 58, 2490–6.
Kuhn, T. S. 1962: *The structure of scientific revolutions* (Chicago).
Lakatos, I. 1970: Falsification and the methodology of scientific research

programmes. In Lakatos, I. and Musgrave, A., editors, *Criticism and the growth of knowledge* (Cambridge) 91–195.

Leighly, J. 1979: Drifting into geography in the twenties. *Annals of the Association of American Geographers* 69, 4–9.

Mackin, J. H. 1963: Rational and empirical methods of investigation in geology. In Albritton, C. C., editor, *The fabric of geology* (Reading) 135–65.

Mackinder, H. J. 1887: On the scope and methods of geography. *Proceedings of the Royal Geographical Society*, new series, 9, 141–60; discussion, 160–74.

Masterman, M. 1970: The nature of a paradigm. In Lakatos, I. and Musgrave, A., editors, *Criticism and the growth of knowledge* (Cambridge) 59–89.

McKenzie, D. P. 1977: Plate tectonics and its relationship to the evolution of ideas in the geological sciences. *Daedalus* 1977(1), 97–124.

Meyer, D. 1973: Urban locational analysis: a paradigm in need of revolution. *Proceedings of the Association of American Geographers* 5, 169–73.

Moffatt, I. 1977: *Paradigm development in geology* (Department of Geography, University of Newcastle upon Tyne, Seminar Paper 33) 1–34.

Müller-Wille, C. F. 1978: The forgotten heritage: Christaller's antecedents. In Berry, B. J. L., editor, *The nature of change in geographical ideas* (DeKalb: Northern Illinois University Press)38–64.

Parsons, J. J. 1979: The later Sauer years. *Annals of the Association of American Geographers* 69, 9–15.

Planck, M. 1948: *Wissenschaftliche Selbstbiographie* (Leipzig).

Pred, A. 1979: The academic past through a time-geographic looking glass. *Annals of the Association of American Geographers* 68, 175–80.

Roe, A. 1963: Personal problems and science. In Taylor, C. W. and Barron, F., editors, *Scientific creativity: its recognition and development* (London) 132–8.

Rogers, E. M. 1962: *Diffusion of innovations* (New York).

Saey, P. 1968: Towards a new paradigm? Methodological appraisal of integrated models in geography. *Bulletin of the Ghana Geographical Association* 13, 51–60 (published 1973).

Sauer, C. O. 1930: Basin and range forms in the Chiricahua area. *University of California Publications in Geography* 3, 339–414.

Spottiswoode, W. 1861: On typical mountain ranges: an application of the calculus of probabilities to physical geography. *Journal of the Royal Geographical Society* 31, 149–54.

Stamp, L. D. 1966: Ten years on. *Transactions of the Institute of British Geographers* 40, 11–20.

Stoddart, D. R. 1975: Kropotkin, Reclus, and 'relevant' geography. *Area* 7, 188–90.

Stoddart, D. R. 1976: *Mackinder: myth and reality in the establishment of British geography* (Abstracts, Symposium K5, XXIII International Geographical Congress, Leningrad).

Stoddart, D. R. 1980: The RGS and the 'New Geography': changing aims and changing roles in nineteenth century science. *Geographical Journal* 146, 190–202.

Taylor, P. J. 1976: An interpretation of the quantification debate in British geography. *Transactions of the Institute of British Geographers*, new series, 1, 129–42.

Tribe, K. 1978: *Land, labour and economic discourse* (London).

Unstead, J. F. 1949: H. J. Mackinder and the New Geography. *Geographical Journal* 113, 47–57.

Vine, F. J. 1977: The continental drift debate. *Nature* 266, 19–22.

6

On People, Paradigms, and 'Progress' in Geography

Anne Buttimer

The notion of 'paradigm' exercises a growing appeal among historians and commentators on geographic thought (Stoddart, 1967; Whitehand, 1970, 1971; Chorley, 1974; Berry, 1974). As with 'peneplain' a few generations ago, it offers an illusion of clarity yet remains sufficiently vague and analytically elusive to occupy our imaginations for a long time.[1] The theory of scientific revolutions has provoked a virtual cacaphony of protest and acclaim which has exposed several latent conflicts and uncertainties in the history and philosophy of science (Kuhn, 1962; Lakatos and Musgrave, 1970). What emerges from this din with resounding clarity, however, is stronger evidence than before that the evolution of scientific ideas cannot be appreciated without a closer scrutiny of their social, ideological, and political milieux.

Kuhn's work also shows repeatedly how captivating the influence of established scientific conventions is, once they have become 'normalized' within research institutions. Indeed disciplinary identity has become increasingly construed in paradigmatic terms, providing a kind of passport to our respective Academies of Science. And geographers everywhere seem to aspire to join that merry throng.

Two obvious challenges for the historian of geographic thought arise from a consideration of paradigms: first, the relationship between geography and 'science', and secondly, the relationship between geographical thought and milieu. As peneplain in the classical Davisian mythology, paradigms too could be characterized in terms of structure, process, and stage. A process called 'normal science' is associated with each paradigm structure at any particular 'stage', and the movement between stages is the function of 'extraordinary science'. If one were to take a liberal approach to time I suppose one could compare the evolution of Davisian landforms with the evolution of scientific ideas. The sibling rivalry of geology and geomorphology might in many ways be compared with geography's continued cloak-and-dagger interaction with systematic sciences.

This may be as far as one can take the analogy. The idea of 'paradigm' has proven most appropriate for describing developments within physical sciences, it fits less comfortably in the story of biological sciences, and finds itself on rocky territory when applied to any field which aims at comprehensive understanding of humanity and milieu. Its appropriateness for studying the history of geography, therefore, depends upon how one views the relationship between geography and scientific disciplines. And on this question we are far from universal consensus. The *ceteris paribus* assumption, for example, which science needs in order to unravel systematically the structure and dynamics of specific 'slices' of reality, has relevance only to particular dimensions of geographic enquiry. It has been suggested that the ultimate synthesis toward which geography as a whole orients itself is a *contextual* rather than a *compositional* one (Törnebohm, 1972; Hägerstrand, 1974b). If this is so, then each conceptual step toward reaching a paradigmatic understanding of the 'composition' of phenomena can spell progress for geography as a whole if the concept can be woven back into an understanding of its human and environmental context.

The purpose of this paper is not to deal with the various conceptual and semantic nuances associated with the term paradigm (see, however, Masterman, 1970) but rather to raise some issues surrounding the appropriateness of this approach to the history of geograhic thought. I shall also suggest an alternative approach which would focus on the life work of particular individuals whose ideas have been significant in shaping the directions of geographic enquiry throughout history.[2] The rationale for this proposal stems from a malaise about contemporary academic approaches to ideas and practice in general and a strongly felt need to explore the reflective and personal dimensions of thought as well as its analytical and empirical ones. At face value the idea of critical reflection upon personal experience may seem very remote from the study of scientific revolutions. Its premises and procedures stand in marked contrast to the objective generalizing stance implicit in the search for paradigms. There is much common ground, however, to be explored via these distinctive paths, and a focus on individual experience should not detract from the search for general patterns. There are surely many routes possible into the history of ideas and if each is carefully signposted, then our collective efforts may produce a more helpful routemap for future explorers.

Biographical accounts of explorers and scientists bear witness to the value of studying an individual's work in the total context of his time and milieu.[3] Within such case studies one finds ample evidence for the overwhelming importance of general intellectual trends and ideological setting. Logically and empirically, a focus on individuals should thus complement the focus on general developmental themes. The issues to which I wish to attract attention probe more deeply than this, however: I wish to raise questions regarding the nature and quality of the *knowledge* whose history we wish to reconstruct, the processes whereby such historical research is to be done, and the effectiveness of our results for contemporary thought and practice within geography. I cannot promise a perspective which will guaranteee either theoretical or revolutionary results, but I do hope to clear the ground for a more objective and personal approach to the history and philosophy of geography.

THE WHAT AND THE HOW IN THE HISTORY OF IDEAS

Reconstructing the history of thought inevitably involves assumptions and inferences regarding the goals of enquiry and the direction of its development. Besides, one's point of view always reflects the circumstances and *Zeitgeist* of the time at which such history is written (see Granö's chapter in this book, figure 32).

In adopting the perspective of 'scientific revolutions' on the history of geographic thought, one implies that geography shares some common goals and directions with conventional sciences. Granö has indicated a degree of parallelism between the ways in which geography and science as a whole have responded to external societal challenges over time (see Granö's chapter in this book, table 3.1).

To appreciate fully the value of this approach, however, one needs to become aware of what types of knowledge are included or excluded, and what types of research are implied.

The structure of scientific revolutions and paradigmatic normal science are premissed upon two complementary scholarly goals:

(1) intellectual mastery over complexity via abstract generalization and theory; and

(2) increasing analytical acuity in conceptual and methodological frameworks and procedures.

Both reveal an attitude toward the relationship between knowledge and experience which Heidegger characterized as *Rechnendes Denken* (Heidegger, 1954, 1971), a style of thought which separates mind and matter, subject and object, and aims at rational understanding and control over reality.

The alternative attitude which he characterizes as *Besinnliches Nachdenken* is one which refuses to separate knowing and being and seeks to unlock the meaning inherent in reality, as far as possible, in its own terms. Contrasting scholarly roles and research procedures follow from these two attitudes: in the former context scholars assume a detached, 'objective' stance on praxis and emphasize products rather than the process of enquiry. In the latter context scholars assume an 'engaged' (immersed) stance; they try to become aware of their own *a priori* ideas and also to elicit self-awareness among the people and 'objects' they study, emphasis resting on the process as well as on the products of enquiry.

I have, of course, grossly oversimplified a complex philosophical issue but it is one which has been overlooked in the debate over Kuhn's work. In fact the entire debate could be adumbrated with the first of these two catagories: the *Rechnendes Denken* so characteristic of Western science generally. Could the same be said about our discourse on the history of geographic thought? If so, how well does the paradigmatic analysis fit the actual record of what geographers have thought, said, and written, and of how they lived and inspired people associated with their careers?

The distinction between these two attitudes toward knowledge and experience should not necessarily imply a separation and I suppose the ultimate challenge for a scholar is to weave these two perspectives in a mutually creative way. It is in this context that one could argue for a closer look at the unique personal experiences of particular geographers: how they were influenced, how they inspired others, what fresh insights they brought to the field as a whole. To unmask the distinctions between 'subjective' and 'objective' accounts of ideas, their genesis and development, is only part of the rationale for this reorientation. Eventually, its aim is to develop a style of scholarly activity which provokes discovery and self-awareness within the whole of *Dasein*. This style may provoke more questions than answers because it endeavours to promote an attitude of listening to experience rather than striving to represent it in terms of *a priori* formulations.

Ideally, of course, a critical appraisal of conventional models and procedures should always precede their application to the analysis of particular problems. It should therefore be seen as a complement rather than a substitute for scientific research. Development of the attitude of *Besinnliches Nachdenken*, like the cultivation of virtue, is probably a life task and example is probably more helpful than words. It has to be discovered and practised by each scholar in his/her own particular circumstances. Who better than the authors of history, then, to reflect upon the evolution of their own ideas, and share their insights with posterity? (MacIver, 1969; Heller, 1918.)

CHARISM AND CONTEXT: IDEAS AND MILIEUX

The style of research adopted in the history of geographic thought and the attitude associated with it inevitably influence the content of what is produced. Is it possible to develop a style of research into the history of our discipline which allows us to tap a richer range of ideas and insight than conventional ones? Building on what Professor Granö has already presented, let me raise some possibilities for weaving the two contrasting approaches to history under three major headings:
(1) geography and 'normal science';
(2) geography and 'extraordinary science'; and
(3) geographic thought and milieu.
Substituting the term 'experience' for the term 'action' in Granö's diagram (see figure 2, chapter 3 of this book), let me use this general framework to illustrate the major points (figure 6.1).

1. *Geography and normal science*

'Normal science', as defined by Kuhn, is perhaps best exemplified in the analytically-oriented disciplines where hypothetico-deductive procedures are used and law-like generalizations are sought. Under specified conditions these procedures may arrive at statements about reality which can have predictive power (Rudner, 1966; Hempel, 1965, 1966; Ryan, 1970). To achieve this kind of certainty, the scope of enquiry has to be limited: specific types of phenomena or process must be isolated from their full contexts so that their own intrinsic dynamism may be examined

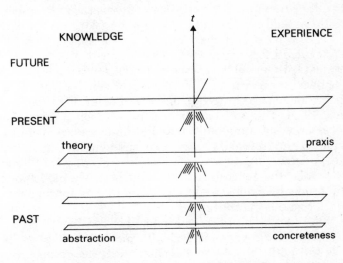

FIGURE 6.1   Reconstruction of geographic thought and practice

more closely. *Ceteris paribus*, an elaborate set of rules and conventions are believed to guarantee a certain type of scientific certainty about carefully circumscribed slices of reality (Popper, 1960, 1962). Normal science excels in yielding insight into the processes whereby things function and inter-act. In this respect its fundamental goal could be described as *technical* as opposed to the *practical* concern for weaving theory and praxis (Habermas, 1971, 1973). With increasing levels of specializaton and compartmental-izaton of scientific effort, the problem of integrating the results of such research becomes an enormous challenge to logic and language. To relate them back into their environmental context is the fundamental problem of contemporary applied sciences – a problem by now familiar to the geog-rapher (Hempel, 1966; Myrdal, 1969).

Critical reflection upon the historical development of the 'normal science' orientation within geography during the past few decades could shed light on these two sets of issues especially if we deliberately keep in mind the relationships between theoretical and practical developments within the field (Granö 1977 and chapter 3 of this book). Plenty of examples could be found in each of our countries where geographic models – spatial, ecological, behavioural – have been applied to social and environmental planning and an evaluation of these experiences could yield a rich variety of human implications (Buttimer, 1974, 1979). Why not begin such evaluation by inviting some of our senior colleagues who have been influential in steering geography toward a more 'scientific' course on the one hand, and toward a more 'relevant' stance on the other, to reflect autobiographically about the evolution of their ideas and praxis? In an atmosphere freed from the pressures of formal academic obligations, perhaps they could now be more free to undertake such a task than are those of us who are surrounded by the constraints of departmental life?

I do not intend to suggest that critical evaluation is not possible via other routes. The work of the Frankfurt School, for example, shows how a hermeneutical approach could derive many general insights into the relationship of science to technological and social life (Marcuse, 1972; Schroyer, 1973). The problem about generalizations, however, is that they necessarily subsume individual and unique ideas within broader rubrics and cannot reveal the total context and import of an individual's thought (Polanyi, 1962; Sorokin, 1957). General critique of trends, too, often has a negative tone: the inappropriateness of a model for general application may be emphasized to the point that its appropriateness for particular situations may be ignored.

Each of the geographic ideas which might be considered as candidates for 'paradigm' status presumably had a moment and place of origin where its 'fit' with the environmental context was optimal. Examples could include the hierarchical networks of central places in Bavaria, ecological zones of land use in early twentieth century Chicago, administrative regionalization schemes based on *pays* in prewar France, or 'behavioural' models of shopping choice in retail sectors of American cities. When applied to other contexts the fit is less comfortable and often quite gauche. To appreciate the positive value of particular ideas and to understand the complexity of applying descriptive models in a normative way one needs to understand the multiple contexts in which these ideas were first articulated and applied (Gellner, 1965; Louch, 1967). Ideally one needs to hear the story from the lips of the authors themselves, many of whom are still alive today.

Strictly speaking, I suppose that wherever geographers have practised 'normal science' enquiry they have worked in conjunction with scholars from other disciplines and shared common models and perspectives with them. In a broader context could we not regard geography as a whole as offering the *practical* challenge to varieties of *technical* expertise, viz., to ask how the results of normal science could be woven back into concrete life circumstances, e.g. in the context of New Towns or regional development policies? (Hägerstrand, 1976; White, 1972). Collectively, too, geographers – to the extent that they remained sensitive to environmental changes – could challenge normal science beyond its established routinized patterns of enquiry.

More interesting than the plateaux of normal science activity then is that ridge-and-valley zone of search for new paradigms which Kuhn labels 'extraordinary science'. It is here that the geographer's unique concern for the environmental appropriateness of ideas could have expressed itself.

2 Geography and 'extraordinary science'

Temporality and change are universal features of life. To avoid the creeping paralysis of routinization and obsolescence normal science must respond to intellectual innovation as well as remain sensitive to changes in the practical and technical challenges of its external milieu (Kuhn, 1962, 1963; Feyerabend, in Lakatos and Musgrave, 1970). In the Kuhnian scheme there is a trial-and-error phase of exploration for new paradigms when the dialectic between conventional scientific ideas and material conditions of life reveals itself. Normal science with all its arsenal of established 'know how' is confronted with questions of 'what', 'wherefore', and 'for whom'. This perspective on the evolution of ideas is certainly a provocative one and dovetails nicely with a Marxist conception of history. The mystery of how knowledge actually develops, however, is still a puzzle to most philosophers of science: not all significant 'breakthrough' ideas can be ascribed exclusively to external challenges (Koestler, 1964; Utah Creativity Research Conference, Taylor, 1964; Ferkiss, 1969). Even in the Kuhnian schema, key insights are almost invariably traceable to the work of 'extraordinary' individuals (Kuhn,

1963; Rank, 1932, 1959; Koestler, 1964). In addition to questions relating to
the dialectic of material and ideological forces within any setting therefore
one should perhaps explore what types of context, what types of interaction, are most conducive toward allowing individuals to express their
creativity (Sorokin, 1957; Rogers, 1962; McMurray, 1972).

An openness of ideological milieu may indeed be among the preconditions for allowing certain types of scholarly interaction to take place
(Ferkiss, 1974; Murphy, 1958).

Toulmin's account of Wittgenstein's Vienna provides an illuminating
example of the context (late nineteenth century) within which several
distinctively new paradigms were born (Toulmin and Janik, 1973). A
better illustration of 'extraordinary science' could scarcely be found anywhere since Ionian times (Barnes, 1965). What this study suggests is that
frequent face-to-face contacts among scholars and between scholars and
people from other walks in life may be the vital facilitators of extraordinary science. A common social challenge and opportunity for informal communication over a considerable period of time may also be
necessary.

Could it not be that the genesis and maturation of great ideas owes as
much if not more to informal *verbal* exchanges between individuals than
formal academic structures or to explicit societal demands? If Koestler is
right many of the great scientists were 'sleepwalking' when their most
creative ideas began to crystallize – they did not even recognize the
profundity of their 'discoveries' until there was an opportunity to share
them with a few friends or associates who helped them to articulate them
(Koestler, 1964; Rank, 1959).

To trace the story of normal science one may look to the records of
research laboratory, professional congress, academic curriculum, or
disciplinary journal. If one wishes to locate the seeds of extraordinary
science perhaps the salon, the field excursion, correspondence file, personal journal or lecture notes might be a more appropriate source. And
for those innovators who are still alive, what better approach than simply
to ask them?

In piecing together the 'objective' history of ideas, we have relied
mostly on documentary evidence. Working from such sources, one can
only make guesses or inferences about the nature or sources of influence
on a scholar's ideas and work. Yet I suspect that there have been pioneers
of geographic thought whose inspiration flowed through their teaching
and field experiences, through their counselling and listening, and in
some cases at least, they did not commit their ideas to print. Should we
not try to elicit some of this story from older colleagues who may still
remember their mentors, some of whom may never have bequeathed a
printed account of their ideas?

Only an autobiographical account could ever do complete justice to the entire story and for many crucially important issues, this may now be impossible. However, once having explored the whole range of any individual's story, one's attitude toward the history of ideas can be profoundly enriched. The challenge would be an interesting one to any of us, at whatever stage of intellectual or professional development we may be. What were the occasions, events, encounters, which have been most influential in encouraging and shaping our thought? If each of us were to reflect upon our own experience and recall what were the critical 'stepping stones' in our thinking, would we not find that particular persons, meetings, letters, lectures, all stand out as occasions of creativity and fresh insight? When communications media and travel facilities were not so easily available, is it not conceivable that the nature and content of interaction among colleagues was quite different? Before the full flowering of the print industry, the number and volume of exchanges may have been smaller but what of the quality?

There has certainly been a trend away from concern about unique places and personalities to a more nomothetic concern for spatial systems and society in general. In our generation, however, there is a strongly felt desire to reverse that trend within ideas as well as within life and to recapture something of the wholeness and centredness offered by community and place. It is not clear, for example, that an exchange of ideas around a seminar table or small gathering of scholars can be a far more useful exercise than reading an array of 'learned' papers? When the 'proceedings' of professional meetings are committed to print, consider how much and what kind of knowledge gets recorded. Does it not involve a summary of whatever thought *preceded* the gathering rather than a record of what participants actually discussed? When one considers how much is inevitably lost when oral discussion is recorded in print one suspects that our documentary record of congresses and textbooks contains only a shadow of the existentially meaningful history of geographic thought.

The expertise required for the pursuit of normal science may be quite different from that which is required for extraordinary science. On this rugged terrain one needs art, ingenuity, courage, and most of all, sensitivity to one's local milieu. These are some of the qualities exemplified in the lives of some older colleagues; why not invite them to author the kind of reflection suggested here? From their shared recollections, we could perhaps not only derive insight into the nature of geography and its history but also gain fresh perspectives on our own routine activities. The structure of academic institutions and the fokways of university life channel our energies along more specialized, print-oriented styles of research and of abstraction. We need to discover how rich is the variety of potential alternatives.

3 *Geographic thought and milieu*

One of the key attractions in Kuhn's theory is the emphasis placed upon milieu – social, psychological, and ideological circumstances surrounding the development of paradigms. To understand the development of geography, however, one needs a broader definition of milieu, even at the risk of shattering the illusion that a theoretical understanding of its influence may ever be attained. Geographic ideas have shaped and been shaped by the physical, technological and economic milieux in which they have enjoyed popularity.

In a wider perspective the history of geography could be regarded as the story of humanity's quest for understanding the earth and arranging it in ways to facilitate its own life interests. If one relies only on documentary evidence, it would not be too difficult to construe it as the record of how certain humans have projected their own self-images on to the earth, calling their projection 'geography' (Santos, 1975a, 1975b; Lefèbvre, 1974). The overwhelming dominance of certain 'classical' themes (e.g. regions, hierarchies, and domains) has been associated with some particular human interests (political, economic, and technological). Others (e.g. the sense of place, 'home', harmony between humans and other life forms within particular places – themes perhaps less amenable to becoming operationalized as 'normal science' or to representation via maps or prose) have become little more than a counterpoint melody. To view the history of geographic thought exclusively in paradigmatic language would almost by definition highlight the *Rechnendes Denken* and diminish the significance of *Besinnliches Nachdenken*. Awareness of the gestalt within a scholar's life and work may be one way of exploring the interpenetration of both in the author's own perception of the process.

The reciprocity of influences between geographic ideas and environmental planning and design has become increasingly apparent in post World War II settings. In this historical fact, one could undoubtedly find support for the Kuhnian approach. There are two vital issues, however, which have not been adequately considered: first, there are many aspects of any milieu – physical, biotic, cultural – which have not yet become so identified with, or transformed by, ongoing scientific paradigms. Secondly, little explicit attention is paid to the local connectedness of phenomena within a particular milieu. These two issues are, of course, intimately related: the specialization and fragmentation of science has been associated with – if not responsible for – increasingly fragmented life styles and milieux. It is perhaps in the context of these unresolved issues that geography could offer a unique contribution to the history of ideas. The homework needed within our own discipline in weaving together our various strands of specialized endeavour could be seen as a microcosm of the problem facing knowledge generally.

The integration of knowledge could quite conceivably be attempted in typical Aristotelian fashion, using the models supplied by mathematics, logic, or other rational systems. Its practical corollary would be managerially designed 'interdisciplinary' research teams or revised school curricula. However rational such an integration might be, it could still remain within the traditional pattern of *Rechnendes Denken* and still imply a faith in programmes rather than persons. Two alternative bases for the quest for integration which this alternative approach offers are (1) the human person, and (2) concrete life situations. Is it vain to hope that if the drive toward conceptual clarity and evolutionary continuity were consistently tempered by concern for concreteness of experience and milieu one could develop a more sensitive and responsible kind of knowledge?

The two foundations on which this hope rests can be justified on experiential rather than logical grounds: first is the assumption that a person *cares* about his own life, and second, that (in some cases at least) he *cares* about his home environment (Rogers, 1962; McMurray, 1972). It is this quality of caring which has been evident in the thought and life of scholars and which may be one of the essential preconditions for a *Besinnliches Nachdenken*. And curiously this is the very quality which is not amenable to articulation via paradigmatic normal science or in the 'official' documentary record.

In the shared autobiographical accounts of geographic thought and experience within specific milieux and specific time periods I suspect one could count on discovering some clues to the solution of the problems of integration and specialization (1) within the minds of particular persons, and (2) within the story of particular regions/places where a geographer's work took place.

TOWARD A MORE INTEGRATED PERSPECTIVE ON GEOGRAPHIC THOUGHT

Geographic ideas at any period can thus be best appreciated if the sociological and ideological influences on their connections with praxis can be unmasked. This much is already amply defended within the paradigmatic approach to the history of science. What is further required for an appreciation of geographic thought is to see the ways in which both ideas and praxis impinge upon and reflect the overall physical and technological milieu. At any period also there may be fundamental disjunctures between the picture yielded via an 'outsider's' survey of conditions and the 'insider's' perspective on his own creative efforts. The challenge outlined in this paper is to somehow discover a framework of enquiry which would permit the historian of geographic thought to relate both perspectives.

Figure 6.2 sketches certain basic dimensions for such enquiry. Its fundamental aim is to point in the direction of a higher level general-

D

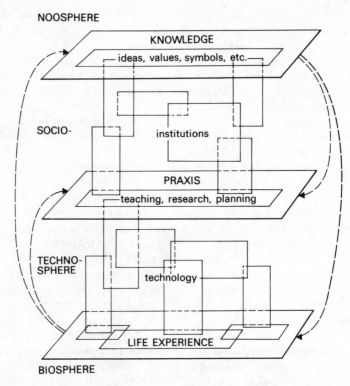

NOOSPHERE

BIOSPHERE

FIGURE 6.2   Cross sectional view of the geographic 'now'

ization regarding the nature and significance of particular developments within geography whether one considers an individual's life work or the import of a particular school. It endeavours to highlight the sociological context of idea formation by underscoring the growing significance of institutions and technology in 'writing the score', as it were, not only within but also among the three levels of enquiry.

To examine the temporal succession of currents within geographic thought and praxis anther diagram (figure 6.3) could elucidate how changes within any one level may reverberate through the others (*ceteris non paribus*). In this way the diagram could be used as a framework within which to test some of the Kuhnian hypotheses. It could also serve as foundation for generalizing the data derived from autobiographical reflections on (1) a scholar's life work, and (2) the life story of a particular region. As with any model this one has both a descriptive and a normative intent: it suggests ideal ways of being as well as of thinking about the world. Essentially it corresponds with classical thinking about the iden-

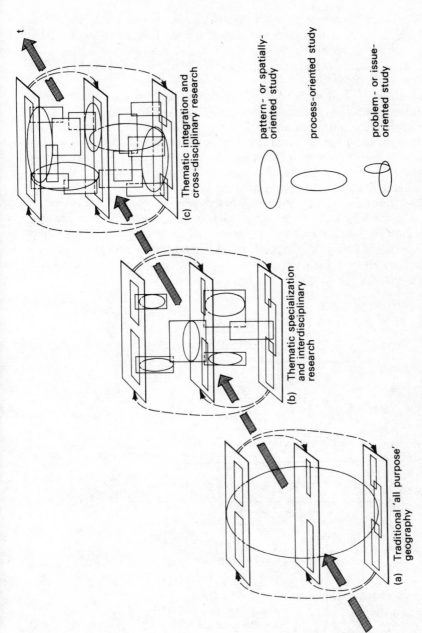

t

pattern- or spatially-
oriented study

process-oriented study

problem- or issue-
oriented study

(c) Thematic integration and
cross-disciplinary research

(b) Thematic specialization
and interdisciplinary
research

(a) Traditional 'all purpose'
geography

FIGURE 6.3   Growth of the socio-technosphere and its impact
upon geographic thought and praxis

tity of geography as a study of the earth as 'home' of mankind. Substituting the term *praxis* with such conventional notions as *functional organization of space/time,* or *genres de vie* (figures 6.4 and 6.5) one could use this diagram to clarify problems relating generally to *genres de vie,* circulation or environmental change.

The implication here is that a geographer could use a model which facilitates an understanding of the 'world' to elucidate the development and significance of his own thought within the discipline. The parallelism between these two axes of enquiry may be more important than is superficially evident. In terms of the diagram, the issue could be described as follows: take any one of the components out of context and develop a specialized line of research and there you find a valuable 'scientific' exercise (figure 6.2). Until the result of this specialized research is woven back into the whole, the full harvest cannot be reaped for humanity and environment. Similarly let any one aspect of personality or talent develop in a specialized way (as, indeed, most 'Promethean' types in Western civilization seem to have done) and unless the results of this specialized

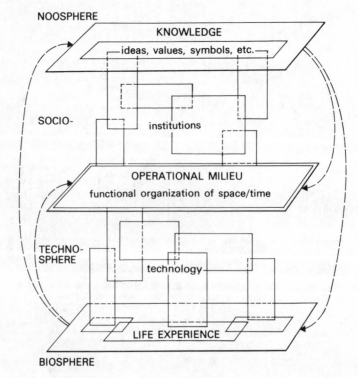

FIGURE 6.4   Structure of geographic enquiry

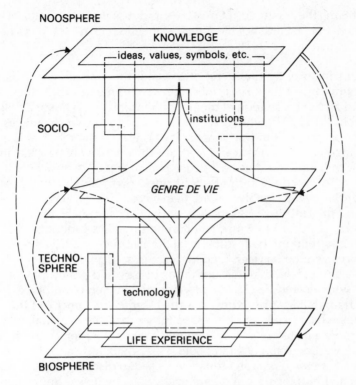

FIGURE 6.5   *Genres de vie* in geographic perspective

effort can be woven into life as a whole, one cannot appreciate its full creative import (Barron, 1963; MacIver, 1969). This is not an argument against specialization of expertise but rather an argument for placing art and science in their appropriate roles within the evolution of life as a whole. To find in one's own life a microcosm of 'world' may be a prospect too challenging to those of us today who have been socialized into a fear of both individualism and fascism but the generation which has never experienced the extremes of either may thirst for some of that wholeness and centredness which can only come by experiencing one's own potential creativity. There is much about the 'spider's web' of administrative protocol and the sometimes stifling paraphernalia of disciplines which needs careful examination.

CONCLUSION: GEOGRAPHY AND CARING FOR KNOWLEDGE

In some queer way things depend for the knowledge of them on our personal interest in them. So soon as we depersonalize our attitude to them, they withold their secrets and dissolve away into sets of general characteristics floating in an ether of abstract intellect. (McMurray, 1972, 192).

Could this not be the secret of the enduring appeal of our pioneers in geography – the *personal* message which emerges from their lives? Beyond all the fascination of method, philosophy, and literary style, it is probably the way in which an author speaks to us and challenges us to become as creative in our contexts as he was in his that has the most enduring appeal. Charism and context – the interplay of personal insight and life milieux – this may be the root explanation of great ideas in human history. To articulate insight about one's own life and milieu at any moment in history involves more than a cognitive mastery of facts from past and present. It is essentially a work of art. For such creativity in the art and science of knowing, emotion and intellect cannot be divorced; feeling and thinking both suffer from the artificially contrived separation which convention has legislated for them in academic life today.

There are indeed many issues in the history and philosophy of geography which could be elucidated via a combination of autobiographical and paradigmatic approaches. It is the essential reciprocity of these two avenues which needs to be stressed rather than their separateness. We need more epistemologically rigorous foundations for research and a more open discussion of values before we can identify a credible core for the discipline of geography today. We need to discover together a set of intellectual horizons for geographic effort as a whole – horizons which would enable us to orchestrate our specialized endeavours in a more harmonious way and yield results which are more carefully integrated and socially responsible.

In addressing ourselves to these concerns, we could find support from colleagues not only in other academic disciplines but also in other walks of life, for example, in medicine, in technology, in theatre and the counselling arts. Perhaps it is time we 'graduated' from our short but intensive flirtation with the idea of being exclusively a social science and rediscover older friends in natural science, music, gardening, and mountain climbing: who knows, we may 'sleepwalk' our way to fresh ideas about humanity's place in earth history and to more life-supporting ideas about contemporary *genres de vie*.

Notes

[1] Comment by Professor Olavi Granö in a discussion about the Edinburgh Symposium, Lund 1976.
[2] 'Proposal Concerning Philosophical Critique and Improved Communication within Geography' presented to the I.G.U. Symposium, Leningrad, 1976.

³ Historians have long recognized the value of oral accounts of events and the autobiographical perspective. See Mason, E. B. and Starr, L. M. editors, *The oral history collection of Columbia university*, Columbia University, New York. In both physics and psychology an autobiographical approach has been initiated. *The History of Psychology in Autobiography* series now has at least six volumes. The American Institute of Physics Center has also refined a format to be used by individuals when writing their autobiographical accounts.

References

Barnes, H. E. 1965: *An intellectual and cultural history of the Western world*, Vol. 1 *From earliest times through the Middle Ages* (New York).

Barron, F. X. 1963: *Creativity and psychological health: origins of personal vitality and creative freedom* (Princeton, N. J.).

Berry, B. J. L. 1974: A paradigm for modern geography. In Chorley, R. J., editor, *Directions in geography* (London) 1–21.

Buttimer, A. 1974: Values in geography. *Association of American Geographers, Resource Paper* 24.

Buttimer, A. 1979: Erewhon or Nowhere Land. In Olsson, G. and Gale, S., editors, *Philosophy in geography* (Dordrecht) 9–37.

Chorley, R.J., editor, 1974: *Directions in geography* (London).

Ferkiss, V. C. 1969: *Technological man: the myth and the reality* (New York).

Ferkiss, V. C. 1974: *The future of technological civilization* (New York).

Gellner, E. 1965: *Thought and change* (London).

Granö, O. 1977: Geography and the problem of the development of science. *Terra* 89, 1–9.

Greer, S. 1969: *The logic of social enquiry* (Chicago).

Habermas, J. 1970: *Toward a rational society* (Boston).

Habermas, J. 1971: *Knowledge and human interests* (Boston).

Habermas, J. 1973: *Theory and practice* (Boston).

Hägerstrand, T. 1974a: The domain of human geography. In Chorley, R. J., editor, *Directions in geography* (London) 67–87.

Hägerstrand, T. 1974b: Studier i samverkans tids- och plats-beroende. *Svensk Geografisk Arsbok* 50, No. 508, 86–94.

Hägerstrand, T. 1976: The geographer's contribution to regional policy: the case of Sweden. In Coppock, J. T. and Sewell, W. R. D., editors, *Spatial dimensions of public policy* (New York) 243–57.

Heidegger, M. 1954: *Vorträge und Aufsätze* (Pfullingen).

Heidegger, M. 1971: *Poetry, language and thought* (New York).

Heller, O. 1918: *Prophets of dissent* (Port Washington, N. Y.)

Hempel, C. 1965: *Aspects of scientific explanation* (Glencoe).

Hempel, C. 1966: *Philosophy of natural science* (Englewood Cliffs, N. J.).

Koestler, A. 1959: *The sleepwalkers* (London).

Koestler, A. 1964: *The act of creation* (New York).

Kuhn, T. S. 1962: *The structure of scientific revolutions* (Chicago).

Kuhn, T. S. 1963: The essential tension: tradition and innovation in scientific research. In Taylor, C. W. and Barron, F., editors, *Scientific creativity: its recognition and development* (New York) 341–54.

Lakatos, I. and Musgrave, A. 1970: *Criticism and the growth of knowledge* (Cambridge).

Lefèbvre, H. 1974: *La production de l'espace* (Paris).
Leiss, W. 1974: *The domination of nature* (Boston).
Louch, A. 1967: *Explanation and human action* (Oxford).
MacIver, B. M. 1969: *Moments of personal discovery* (port Washington, N. Y.).
Marcuse, H. 1972: *Studies in critical philosophy* (Boston).
Masterman, M. 1970: The nature of a paradigm. In Lakatos, I. and Musgrave, A., editors, *Criticism and the growth of knowledge* (Cambridge) 59–89.
McMurray, J. 1972: *Reason and emotion* (London).
Murphy, G. 1958: *Human potentialities* (New York).
Myrdal, G. 1969: *Objectivity in social research* (New York).
Polanyi, M. 1962: *Personal knowledge. Towards a post-critical philosophy* (Chicago).
Popper, K. 1960: *The poverty of historicism* (London).
Popper, K. 1962: *The open society and its enemies* (London).
Popper, K. 1970: Normal science and its dangers. In Lakatos, I. and Musgrave, A., editors, *Criticism and the growth of knowledge* (London) 51–58.
Rank, O. 1932: *Art and artist. Creative urge and personality development.* Translated by Atkinson, C. F. (New York).
Rank, O. 1959: *The myth of the birth of the hero and other writings,* Freund, P., editor, (New York).
Rogers, Carl R. 1962: Toward a theory of creativity. In Parnes, S. J. and Hardney, H. F., editors, *A Source Book for Creative Thinking* (New York) 63–72.
Rudner, R. 1966: *Philosophy of social science* (Englewood Cliffs, N. J.).
Ryan, A. 1970: *The philosophy of the social science* (London).
Santos, M. 1975a: *L'espace partagé* (Paris).
Santos, M. 1975b: The periphery at the pole: the case of Lima. In Gappert, G. and Rose, H., editors, *The social economy of cities. Urban Affairs Annual Reviews, 9,* 335–60.
Schroyer, T. 1973: *The critique of domination. The origins and development of critical theory* (Boston).
Singh, R. L. 1977: Role of geography as an integrating science. *National Geographical Society of India: Research Bulletin* 23, 1–22.
Sorokin, P. A. 1957: *The crisis of our age: the social and cultural outlook* (New York).
Stoddart, D. R. 1967: Growth and structure of geography. *Transactions of the Institute of British Geographers,* 41, 1–14.
Törnebohm, H. 1972: *Perspektiv pa studier. Avdelningen för vetenskapsteori* (Göteburg University), report no. 41.
Toulmin, S. and Janik, A. 1973: *Wittgenstein's Vienna* (London).
Utah Creativity Research Conference 1964: Taylor, W., editor, *Widening horizons in creativity,* Proceedings of the Fifth Conference (New York) 112–21.
White, G. 1972: Geography and public policy. *Professional Geographer,* 24, 101–4.
Whitehand, J. W. R. 1970: Innovation diffusion as an academic discipline: the case of the 'new' geography *Area* II, No. 3, 19–30.
Whitehand, J. W. R. 1971: In-words outwards: the diffusion of the 'new' geography, 1968–70 *Area* 3, No. 3, 158–63.

7

Wilhelm Dilthey's Philosophy of Historical Understanding A Neglected Heritage of Contemporary Humanistic Geography

Courtice Rose

It is generally agreed among historians of geography that the contemporary form of the discipline represents several major traditions, among which several thematic thrusts are discernible:

(1) geography as the study of areal differentiation, a definition derivative of the Kantian classification of sciences (1765) and later expounded by Alfred Hettner and Richard Hartshorne;

(2) a 'géographie humaine' tradition, also a neo-Kantian formulation which treats the larger sphere of the relationships between man and his environment and is derivative of the works of Friedrich Ratzel, Frédéric Le Play, Camille Vallaux and Paul Vidal de la Blache;

(3) a 'geography as spatial science' tradition, derivative of the 'exceptionalist' portion of the chorological tradition; and

(4) an ecological orientation derivative of the 'man-land' studies of Harlan Barrows, Carl Sauer and J. K. Wright (Hartshorne, 1939; Taylor, 1951; Glacken, 1967; Fischer et al., 1967; Dickinson, 1969; James, 1972. It would be a mistake to assume that these various traditions were unrelated: both the areal differentiation and the 'géographie humaine'

traditions were concerned with finding useful knowledge either in the earth's physical patterns and processes or in the interaction between man and those physical patterns. These two traditions have not remained unchanged in the twentieth century. The largely positivistic tradition, of which Hettner and Hartshorne were the antecedents, has been given a stronger scientific rationale through the development and use of quantitative methods, model-building and theory construction (Schaefer, 1953; Bunge, 1966; Chorley and Haggett, 1967; Haggett, 1965; Harvey, 1969). On the other hand, the 'géographie humaine' tradition has experienced a slower development and has become somewhat fragmented in its scope. The contemporary expression of this school is now best viewed as an amalgam of the ecological orientation with the behaviour and perception schools of Kates, Burton, White, Wolpert, Downs (Saarinen, 1969; Downs, 1970) and the newly emerged group of humanistic and social geographers: Lowenthal, Relph, Buttimer and Yi-Fu Tuan.

It seems clear that twentieth century geographers have, for the most part, ignored the philosophies of Hegelian idealism, Marxism and existentialism (Samuels, 1971), having instead opted for positivism and particularly British analytic philosophy (Harvey, 1969). And although the 'géographie humaine' tradition owes a great deal to Vidal de la Blache and others of the French 'regional' tradition, this debt stretches perhaps further than some commentators would have it. Few, if any, of geography's historians have discussed the role of the German life-philosophers, and Wilhelm Dilthey in particular, as the possible antecedents of the present humanistic tradition even though their influence is implicit in the works of Sauer, Wright, Lowenthal and Spate.[1] Very broadly, the work of the German life philosophers was rooted in the eighteenth-century protests against scientism, formalism and the more 'antiseptic' forms of rationalism which were prevalent in that era, and was an attempt to return to the fullness of lived experience as the basis for a philosophy of man (Palmer, 1969; Bollnow, 1958). Part of the reason for this neglect could be that for geographers such as von Richthofen, Ratzel and Hettner, the discipline was properly part of *Erdkunde* (general earth science), or *Länderkunde* (general knowledge of regions and countries) and not part of the *Geisteswissenschaften*, the systematic human sciences, or the *Kulturwissenschaften*, the cultural sciences. It is certain that several German geographers including Hettner, A. Penck, Schlüter, Banse and Kraft knew of the work of the neo-Kantian philosophers Wilhelm Windelband and Heinrich Rickert and so possibly of the works of Wilhelm Dilthey. Hettner discusses the approach of the neo-Kantians, as well as the terms *Geisteswissenschaften* (the systematic human studies) and *Zusammenschau* (instinctive conception or unified vision), both favourite topics of Dilthey (Hettner 1927, 1929). In ignoring the traditions of the German life-philosophers, among whom are to be found Marx, Nietzsche, Dilthey,

James, Dewey, Scheler, Pestalozzi, Simmel and Weber, a valuable connection may be lost between the historical roots of modern social science
and the emerging humanistic movement in geography. Of these philosophers, Dilthey can be considered as a key antecedent to contemporary
humanistic geography.

1 WILHELM DILTHEY AND THE 'CRITIQUE OF HISTORICAL REASON'

Wilhelm Dilthey (1833–1911) was probably the first thinker to merit the
title 'philosopher of the human studies'. A genuine eclectic, Dilthey was a
powerful blend of biographer, historian, artist, musician, poet, philosopher and teacher. His work was inspired by several currents of
thought: the British empiricism of Locke and Hume, the French realism of
Auguste Comte, the post-Kantian transcendental philosophy of Fichte,
Schelling and Hegel, the romantic humanism of Goethe and Schiller and
the historicist movement of Ranke, Carlyle and Trendelenburg. A contemporary of Marx and Nietzsche and a good acquaintance of both
William James and Edmund Husserl, Dilthey stands almost alone between neo-Kantian and nineteenth century German idealist thought and
twentieth century continental philosophy. His work can be viewed as
directly introductory to the modern existentialism of Jaspers, Buber,
Spranger and Bollnow, to the phenomenology of Husserl and Schutz,
and to the idealism of Croce and Collingwood. The influence of Dilthey's
work can also be seen in psychology (James, Wundt) and in sociology
(Simmel, Troeltsch, Mannheim and Weber). Often classified as 'rela-
tivistic' on 'pluralistic,' Dilthey's work was left largely unfinished and
while it showed some similarities to that of other life-philosophers, particularly I. A. Richards, it moves largely among mutually hostile traditions
and attempts to cover genuinely unknown territory. Dilthey was decidedly Kantian in his earlier works and almost pre-existentialist and
pre-hermeneutic in his later efforts.[2] Hence the diverse themes which
characterize his work – it can be viewed at once as a contribution to
psychology, history, philosophy or hermeneutics.

Dilthey's lifelong task was, as he termed it, the writing of a 'Critique of
Historical Reason'. If Kant's purpose in writing the *Critique of Pure Reason*
was to ask how scientific knowledge was possible, then Dilthey's
announced purpose was to address this same question to the contents of
the mind: how can we explain the moral, aesthetic, religious and social
elements of human events? Rejecting the 'phenomena-noumena' distinction of Kantian epistemology, Dilthey suggested that lived experience or
Erlebnis was the basic unit of consciousness in the human studies.[3] *Erlebnis*
was not indentical to experience in general (*Erfahrung*) or any 'conceptual
ordering of inert sensations' (as Kant had used the term). *Erlebnis*, according to Dilthey, was 'real' and 'finite' in the sense that, with respect to

consciousness, there is nothing hidden behind lived experience as there was in the Kantian formulation.

For Dilthey, the study of lived experience meant the study of 'the whole man', and any philosophy of man derivative of this study must go beyond the realm of simple ideas and concepts. In 1883, he wrote:

> In the veins of the knowing subject constructed by Locke, Hume, and Kant runs no real blood, but the diluted fluid of reason in the sense of mere thought-activity. But I was led, by my concern as historian and psychologist with the whole man, to make this whole man in the full diversity of his powers, this willing, feeling, thinking being the foundation for explaining even knowledge and its concepts (such as those of the external world, time, substance, cause), however much it may seem that knowledge weaves these its concepts only from the material of perception, imagination and thought. (From Preface to the *Introduction of the Human Studies, Gesammelte Schriften*, vol. I (1922), xv–xviii)[4]

Consciousness then, was not merely rational, it must concern feelings, emotions, volitions and all other aspects of experience. Moreover, it cannot be isolated from the rest of a person's being; it is a totality. Whereas Kant's concept of consciousness had been 'pure' or 'transcendental', Dilthey held that the conscious component of lived experience was a feature of this very world and did not participate in any transcendental realm.[5] The task then, was to experience this temporal flow of consciousness and describe its structure. To do this Dilthey began with analytical psychology and the study of aesthetics and ended some twenty-five years later with an attempted hermeneutic of lived experience (Dilthey, 1900, 1906).

2 THE *GEISTESWISSENSCHAFTEN* (HUMAN STUDIES)

For Dilthey, the *Geisteswissenschaften* were 'the totality of the sciences which have historico-societal reality as their subject matter' (Holborn, 1950, 98) or simply, the human studies.[6] In this group he placed psychology, anthropology, political economy, law, history, aesthetics and philosophy. These disciplines were all conceived as interrelated but relatively independent; there was to be no one system of the human studies. The *Geisteswissenschaften* had to be empirical, in part, in order to avoid any metaphysical speculations and yet they must deal with the peculiar subject matters of experience, consciousness, inner perception, cognition, feeling and volition. On the other hand, the human studies must not be solely introspective or involved with self-observation only. Otherwise the human studies could not claim to understand the intersubjective (social) world and could not interpret the 'life-expressions' found in the world, i.e. concepts, judgements, human actions and emotive expressions such as facial expressions, gestures and sighs.[7] Mental life must somehow be 'objectified' if it was to be studied in depth. For Dilthey, this objectification was accomplished by the fact that all inner experience possesses an initial intelligibility. At this level, there was no need for an epistemology to tell us what is the 'test' of our knowledge as in scientific pursuits, our own acts of consciousness are sufficient tests for our knowledge, e.g. 'I exist', 'the world exists', 'this other person exists and is in my office', etc. Using this reasoning all historical understanding in the human studies presupposes a kind of reflective experience which, at once, 'recognizes any historical or natural product as having a possible human significance' (Makkreel, 1975, 222). Dilthey thus considered that 'reflection' was part of all psychic experience and therefore the inner experience of such reflection could be made the object of legitimate inquiry.

In the natural sciences, Dilthey argued, the importance of perception had led to a situation in which human events were modelled after physical events. The human studies, however, require that man, if he is to be studied at all, must be considered in the state in which he is found, that is, in such states which are lived, expressed and understood by other people.[8] The crux of Dilthey's message was: 'the natural sciences we explain, the human sciences we understand' (*Gesammelte Schriften V*, 144). Following Ranke and Schleiermacher, Dilthey asserted that the differences between the natural sciences and the human studies were differences of content, as well as of method. The natural sciences are characterized by the absence of reference to human experience, the human studies, on the other hand, inevitably must deal with all of human experience including the objects of the natural sciences. There is no unique way of knowing characteristic of the human studies, the essential difference lies in the fact that the context of the human studies is a much wider one including, as it does, the contents of experiences in both the inner and outer realms. Accordingly, the human studies have no absolute domain nor are the relationships between its 'objects', i.e. the 'bits' of lived experience, constant in any way – the human studies are developmental in nature rather than constructionist.

If 'the proper study of mankind is man', then the human studies must focus on the individual's consciousness; it must preserve his individuality, understand his context and above all recognize his purposes. To this end the human studies are to be marked by a certain 'purposiveness' in which all concepts are automatically axiological, selective and value-laden[9]. It followed that the discovery of laws similar to those found in the natural sciences was not possible; rather what could be stated were empirical generalizations or tendencies which would not be additive or

hierarchical but separate and complementary in nature. Dilthey's departure from the Hegelians in his conception of the human studies rests on
the fact that he refused to treat any of the factors of mental life in isolation.
Fichte's notion of 'will' and Schelling and Hegel's 'reason' were, it was
true, partial studies of the individual consciousness, but Dilthey argued
that constructs such as 'cognition', 'feeling' and 'volition' are never
experienced separately, they are simply different parts of a whole self, an
integration of all the powers of consciousness (Hodges, 1944, 99).

In 1895, Dilthey also parted company with the neo-Kantians Windelband and Rickert, over the nature and scope of the human studies.
Windelband had claimed a year earlier that historical knowledge had no
intrinsic relation to psychic experience. The former was clearly an object
in the 'outer' world, the latter a creature of the 'inner' world. Dilthey's
answer is in part drawn from the *Critique of Judgement:* psychological
experience was neither a feature of the inner or outer worlds but rather a
'reflective' or 'transcendental' experience.[10] Since when we reflect on our
own inner experiences (to 'objectify' them), we must understand the self
'from without', i.e. understand the self as an object, we can also in a
similar fashion enter into the minds of others and understand 'reflectively' their thoughts. Thus these cultural 'objects', the other people, can
be placed within the grasp of our own reflective experience. Understanding others therefore presupposes self-understanding and thus the only
way in which an objective understanding of history could be obtained
was by 'making our inner life universal' (*G. S.* V, 281). The human studies
were then to focus on the 'individualizing' of cultural concepts, understanding them reflectively as an individual person would, whereas the
cultural sciences (of the neo-Kantians) used an essentially 'individuating'
methodology concentrating on how the isolated individual fits into an
external social, political or economic scheme (Makkreel, 1969, 439).

3 *ERLEBNIS* (LIVED EXPERIENCE)

Dilthey's concept of the human studies is based on his notion of *Erlebnis*
or lived experience.[11] The real units of mental life, he stated, were 'total
reactions of the whole self to a situation once confronted' (Hodges, 1944,
43). In this original act of lived experience there is no subject-object
separation: 'In the act of living experience, perception and the contents of
my perception are one' (*G. S.* VII, 27). As such, lived experience is not
simply one act but a complex of acts. There were, according to Dilthey,
three main components of lived experience: the cognitive, the affective
and the conative. Every unit of action contained all three components and
they occurred in the same order within the unit of action.[12] The relationship between these three elements constituted a structure of the mind, a
structure which was not discovered by inference or by hypothetical

methods but only by experience. The description of this structure was the task of psychology. As such the concept of lived experience represents a considerable improvement over the term *Erfahrung* (the German word for experience in general) and for Dilthey it included *Innewerden* (inner perception or apprehending) as well. *Erlebnis* does not refer simply to the 'content' of the act of consciousness, for then it would be something of which we could be conscious but it is the act itself – it is experiencing as it is given[13] (Palmer, 1969). *Erlebnis* can subsequently become an object for our reflection such as when we (now) reflect on what we did last evening, but the essential character of *Erlebnis* is 'that which in the stream of time forms a unity in the present because it has a unitary meaning' (G. S. VII, 194). Thus while we normally find our perception of people and things accompanied in quick succession by feelings about these people or things and perhaps by determinations to do something about them, these successive stages of our consciousness simply belong together for us and 'we see ourselves as it were out of the corner of our eye while our attention is fixed on something else' (Hodges, 1944, 44). Their unity is a functional one and not unnatural in any way, e.g. the meaning of fear for a friend's illness cannot be readily divorced from our knowing about that condition, our feelings for the friend and then our desire to help if possible.

The relationship of perception to lived experience is an important one for Dilthey. If lived experience exists before any subject-object separation and is an act of consciousness, then it cannot be interpreted as something which 'consciousness stands over against and apprehends' (Palmer, 1969, 108). Experience then, cannot directly perceive itself. This meant that in the structure of the mind there must exist at least two types of relationships, one set which is experienced and another set which is inferred. The experienced relationships were called 'orderings' (*Anordung*) in which 'psychic phenomena are related by inner connectedness; every such datum so connected is part of a structural nexus' (G. S. VII, 15). These orderings then, were 'part-whole' relationships. The inferred relationships were called 'regularities' which were taken out of the nexus of psychic process and observed at a distance, either hypothetically or experimentally. Perception was one of these latter relationships: statements could be made, by way of generalizations about these regularities since they did conform to rules, but these statements were only applicable to the external world. Thus the complex of acts included in lived experience did contain identifiable perceptual regularities. But *Erlebnis* was a more inclusive notion wherein these regularities were often inseparable from the part-whole relationships. Only when the lived experience was intentionally reflected upon was it possible to distinguish clearly between these two types of relationships.[14]

Lived experience also has an important temporal structure. Since lived experience is not static in any way, it tends to encompass past acts,

present thoughts about those past acts, present acts, feelings and volitions as well as future (intended) acts. This temporal context forms the
core of meaning-giving for Dilthey. Any incident or experience in one's
life depends in some way on other (past) incidents and at the same time, it
opens the way for future incidents: meaning is contained along these
linkages between the temporally separated components of an incident.[15]
Meaningfulness is related to the unity or lack of unity manifested by these
linkages. Thus temporality is not something which is superimposed on
experience by a reflecting consciousness, temporality is already explicit in
experience itself. Dilthey describes what happens when an experience
becomes the object of his own reflection:

> I lie awake at night worrying over the possibility of completing in my
> old age the work I have begun; I think over what to do. There is in
> this experience a structural set of relationships: an objective grasp-
> ing of the situation forms the basis of it, and on this is based a stance
> as concern-towards and pain-over-the-objectively-grasped fact
> along with a striving to go beyond the fact. And all this is there-for-
> me in this it's (the fact's) structural context. (G. S. VII, 139)

4 THE IDEA OF *VERSTEHEN*

As with his concept of lived experience, Dilthey's concept of *Verstehen*, or
understanding, also went beyond the purely intellectual and cognitive
processes of the mind. If we were simply involved in explaining certain
events, Dilthey argues, this could be achieved merely by manipulating
several of our cognitive powers such as the capacity for mathematical
symbolization, the operations of logic and the ability to conceptualize
both tangible and intangible objects. But understanding must involve *all*
of the mental powers we have – it is the ability we possess both to
comprehend ourselves and to grasp the mind of another person.[16] Dilthey
provided a certain fullness to the concept of *Verstehen* whereby we come
to know ourselves and others by that special capacity of the mind to grasp
the presence of life itself: life in one person therefore 'understands' life in
another: 'understanding is a rediscovery of the I in the thou' (G. S. VII,
191).[17]

Verstehen thus opens to us our own world as well as that of another
individual. But *Verstehen* was not to be confused with 'empathy'. Dilthey
regarded any empathetic projection of the self into the other as an obstacle
to understanding since empathetic projection could induce us to read in
our own concerns to our understanding of others, e.g. we can understand
the drama portrayed by stage actors without empathizing with them.[18]
Neither was *Verstehen* to be confused with introspection or solopsistic
modes of thinking. That it could not be was the result of two assertions.

First, *Verstehen* presupposed some form of lived experience. Dilthey held that if the validity of any interpretation of a psychic experience is rooted in that experience itself, then a purely introspective or 'inner' experience would be open to our distrust since, again, one could easily fabricate additional portions of this experience, thereby invalidating it as a basis for communication. But in reflective experience, *Erlebnis* is expanded to contain an external reference by virtue of the fact that the experience is being considered at some point in time *other* than when it originally occurred. Hence *Verstehen* can be viewed as an extension of *Erlebnis*. Second, there can be an intermediary in the *Erlebnis-Verstehen* connection, i.e. expression (*Ausdruck*). In any of our reflective thinking, we can only understand ourselves *from without*, just as we observe others. This possibility of reflecting on oneself as an 'historical entity' is facilitated by some expression which intercedes between our having the experience and our reflecting upon that experience. This expression could take several forms: those which convey ideas, e.g. symbols, signs, signals, bells, words, etc.; those which convey human actions, e.g. arm-raising, leg-moving, head-shaking; and those which convey life expressions, e.g. sighs, laughs, gestures, attitudes, etc. The purpose of the expression is to act in a mediating role between the inner and outer realms, or between the self and the other, as a text for the operation of *Verstehen*, thus: *Erlebnis* – expression(s) – *Verstehen*. Once the inner experience has been expressed in some outer form, the interpretation of that expression could procede with *Verstehen*, e.g. the expression of one of my past acts might be the thought of the crumpled bumper on my car. This expression, once grasped fully, renders my experience of slipping off the icy road into a ditch in a context wherein I can understand my attendent fears of personal injury as well as my later disgust with my driving ability. This same type of operation takes place when we reflect on another person's consciousness: to relive his experience (*Nacherleben*) you are guided by the expressions given. In doing so you internalize an 'experience-as-experienced-by-him' in exactly the same manner as you grasped reflectively your own experience. This is possible since lived experience takes place in an intersubjective context: we understand another person because we take him to be possessed of an inner life essentially the same as our own. Dilthey thus held that to understand another person's actions meant to relive those actions 'reflectively'. [19]

In a well known example of the operation of *Verstehen*, we cannot (at this point in time) understand the assassination of Caesar in the sense of 're-experience again the act itself', but we can understand the assassination as an historical event in the sense that we can (now) relive the motives and ideas that took place (then) in that historical event called 'the assassination of Caesar' (Tuttle, 1969). This is possible since the operation of *Verstehen* allows us access to the other's personal experience: the

situation of conflict arising between Brutus and Caesar, the resulting
jealousies and fears, the motives for social reward, etc. This type of
knowledge does not search out strict cause/effect relationships between
elements of consciousness but rather elucidates 'functional, two-way
relationships, relationships which are embedded in the complex of acts of
lived experience itself. This type of knowledge stands in sharp contrast to
the type of knowledge obtained by searching empirical data for its
'coherence' or 'fit' with the real world.

5 FROM DILTHEY TO HERMENEUTICS

Notwithstanding the importance of the earlier works of Schleiermacher
as the basis of modern hermeneutics,[20] it can be stated that Dilthey's
special relationship with Husserl and his resulting hermeneutic approach
to the human studies mark important linkages with the phenomenology
of both Husserl and Schutz. After their first meeting in 1905, Dilthey
corresponded with Husserl for a period of nearly six years. The development of Dilthey's ideas during this period involved both a theoretical
framework for the human studies based in phenomenology as well as the
increasing importance of interpretation as an essential ingredient in any
methodology for the human studies.

Husserl's 'Logical Investigations' (1901) had a pervasive influence on
Dilthey's thinking. Between 1905 and 1909 Dilthey's papers delivered to
the Prussian Academy indicate that his previous notion of lived experience with its elementary 'logical' operation, was actually confirmed by
Husserl's phenomenological account of the prediscursive operations of
consciousness (Makkreel, 1975, 290). Both Husserl and Dilthey had replaced the naive conception of 'inner experience' with a more fundamental notion of 'structural relations in a psychic nexus'. And both had
similar objectives: a description and analysis of psychic structure which
would deliver a firm epistemological foundation for the human studies
without resorting to metaphysical abstractions. Both men agreed that a
phenomenological description of *Erlebnis* was the proper route to this
objective, yet there were to be some diverging views on historicism, on
the possibility of the self-transcendence of lived experience and on the
re-evaluation of dynamic psychology.

Whereas Dilthey could easily see the value of phenomenological analysis
for an epistemology of the human studies, Husserl saw phenomenology
as a basis for all sciences (Husserl, 1911). Dilthey was to insist throughout
the latter years of his life that the historical dimension of human study
could never be reduced to its 'natural', i.e. physical, origin and that
therefore the objectivity of human studies could only be preserved by
keeping its methods separate from the natural sciences. Husserl, on the
other hand, claimed that history offered no clues to the questions of

validity of hypotheses or to the absolute goals of science and that if followed rigorously, Dilthey's use of historical consciousness would ultimately carry over into an extreme subjectivism.[21]

Concerning the self-transcendental character of lived experience, Husserl moved on from the position in the 'Logical Investigations' to an increasingly idealistic form of phenomenology in the 'Cartesian Meditations' wherein general 'essences' are constituted in and by the transcendental ego. But for Dilthey, *Erlebnis* meant the definition of inner experience in terms of its natural relations to outer experience. As such it operates 'totally within the presuppositions of empirical consciousness' (*G. S.* VII, 12). Objects in the external world remain presupposed but there is no way of affirming their reality. Hence there is no phenomenological 'bracketing' of the world; it simply is there. Dilthey's 1905 discussion of the nature of feelings indicates his aversion to the use of the Husserlian époché: 'A feeling exists insofar as it is felt, and it is the way it is felt: the consciousness of it and its nature, its being given and its reality are not separable' (*G. S.* VII, 27).[22] Apprehension of objects in lived experience is different from their apprehension through the senses. Perception, according to Dilthey, was always inherently perspectival and thus incapable of the all-encompassing apprehension necessary for the understanding of *Erlebnis*. In this distinction between 'psychic' objectivity and 'natural' objectivity, Dilthey had implicitly begun to use Husserl's theory of intentionality.

Having added 'expression' as an intermediary between *Erlebnis* and *Verstehen*, Dilthey's turn to hermeneutics forced him to reconsider his programme for a structural psychology. By 1910, Dilthey was claiming that the hermeneutical method was more appropriate for the human studies because it focused particularly on the *Erlebnis-Ausdruck-Verstehen* process and because it contrasts the apprehension of the psychic nexus with the intuitive perceptual methods of the natural sciences (Makkreel, 1975, 300). In this sense his proposed methodology for the human studies is much more dynamic in nature than when it was first proposed in 1883. Specifically, the whole experience is defined from its parts (the meaning relations) and reciprocally, the parts can only be understood in reference to the whole (Palmer, 1969, 118). Just as the meanings of individual words yield an understanding of a whole sentence, so also does this part-whole relationship exist between the parts and whole of one's life. If the sense of the whole thus determines the function and meaning of the parts, then meaning is always something historical, i.e. 'it is a relationship of whole to parts seen by us from a given standpoint, at a given time, for a given combination of parts' (Palmer, 1969, 118). This given combination of parts is a text which has arisen out of the combination of lived experience and its expression (*Erlebnisausdruck*) and meaning and meaningfulness are thus contextual; they are always part of the situation and not something 'outside' the situation which we partially objectify when we render a single meaning explicit.[23] Moreover, meanings can change as the part-whole relationship changes. From this hermeneutic perspective on the methodology of the human studies, all the human studies would be interpretive and mutually dependent, therefore psychology could no longer be held to be primary. Consequently, if the hermeneutic method was followed, there is no fixed starting point for understanding since every part presupposes the existence of the other parts. This means that there can be no 'presuppositionless understanding' – everything to be understood must have some frame of reference – we understand then, by constant reference to our experience. But since experience can also be examined by analytic modes of thinking as well as description, then the methodology of the human studies must incorporate both scientific analysis and hermeneutical description and the original programme for a descriptive psychology becomes a more logical comparative programme (Makkreel, 1975, 303).

6 DILTHEY'S THEMES AND THE STUDY OF HUMAN GEOGRAPHY

Although Dilthey's philosophy of lived experience does merit the position of precursor to the later developments in phenomenology, existentialism and social philosophy, it has, of course, come under close scrutiny and criticism from diverse quarters (Croce, 1921; Troeltsch, 1922; Ryle, 1949; Hodges, 1952; Nagel, 1953; Abel, 1948; Anscombe, 1957; Tuttle, 1969). Even if the 'Critique of Historical Reason' were to have been finished, it is likely that it would not have lived up to its announced purpose of stating an epistemology for the human sciences.

The impact of Dilthey's philosophy of historical understanding on geographers who were his contemporaries is difficult to assess. At best he received only a passing glance from the major methodologists Ratzel and Hettner and while he seems to have been indirectly evident in the *Landschaftskunde* school of geography (Schlüter, Passarge, Waibel, Schmieder, Lautensach and Banse), Dilthey's ideas have not been directly applied to the philosophical underpinnings of human geography. Yet today, the major themes of Dilthey's works form a permanent backdrop for the playing out of methodological controversies in every contemporary social science and they can be recognized in the geographical literature just as they have been so clearly detailed for psychology (James, Wundt, Maslow, Buber) and sociology (Simmel, Mannheim and Weber).

Ritter

Some 15 years before Dilthey was born, Carl Ritter's publication of the first two volumes of 'Die Erdkunde' indicated that he had already been disturbed by some of the same of the same questions which were later to be considered by Dilthey. Geography, said Ritter, should be empirical; it should progress from observation to observation in the search for general laws (James, 1972). Yet geography is also to be conceived as the study of the earth as the 'home of man' and for such purpose some understanding of the interconnections and causal interrelations between things on the earth's surface must be sought. He expressed this concept in the idea of 'unity in diversity' and frequently used the word *Zusammenhang* or 'internal connection' to refer to this cohesive quality of diverse things within an area or region. This was similar in form (if not content) to Dilthey's conception of human experience – the notion that there is an 'inner connectedness' contained in all events of human experience and that it was the specific task of the *Geisteswissenschaften* to theorize about such connectedness. Further, Ritter's concept of 'wholes' to be found in nature, finds its correct ancestry in the idealist themes of J. R. Forster, Zeune, Schelling, Fichte and Hegel, a tradition with which Dilthey was closely connected (Tatham, 1951). The idea that nature was not simply an empirical entity but rather was developing toward some end or goal was fully acceptable to Ritter and would not have been out of place in the later works of Dilthey.[24] Here Dilthey conceives of *Erlebnis* as 'objective spirit', or that plurality of objectifications that can be empirically discovered through the study of history. The objective spirit was to be articulated through the study of dynamic systems (*Wirkungszusammenhang*) found in historical life and specifically the search for a central structure related to the overall functioning of the system (Makkreel, 1975, 314).[25] It was the task of the *Geisteswissenschaften* to study these dynamic systems and to specify the structure of psychic life. Thus although Ritter's concept of *Zusammenhang* was predicated upon physical entities and Dilthey's notion of *Wirkungszusammenhang* dealt principally with psychological phenomena, there was a recognition in Ritter's work of the tension between empirical study and 'historical' study that was to surface so completely in the works of Dilthey.

Hettner

The connections between Dilthey's works and the ideas of Alfred Hettner, a geographer trained in philosophy, are instructive if not conclusive in their similarity. Certainly Hettner was quite clear that the activities of man *per se* should be excluded from the study of geography. He states that geography does not concern 'the details of constitution and administration of states, the organization of the economic, social and spiritual life, the individual products of art, literature and science' (Hettner, 1927, 18). These could be developed anywhere and only insofar as these affected the general sciences of the earth (*Erdkunde*) or the science of the surface of the earth according to its local differences in the continents, countries, landscapes, regions or localities (*Länderkunde*) should they be part of the discipline of geography. Such other historical and cultural facets of study were properly part of the systematic cultural sciences (*Volkerkunde*) for Hettner and they included: linguistics, religion, political science, economics, and knowledge of nations. But if Hettner would not have agreed with Dilthey that human geography was a part of the *Geisteswissenschaften*, he did, on the other hand, defend a Diltheyian position with respect to the distinctive methodologies employed in the natural and the cultural sciences. Hettner vigorously opposed Windelband and Rickert's contention that the *Geisteswissenschaften* were essentially idiographic in nature, intentionally directed toward producing a full and exhaustive representation of a simple, non-recurring event and limited in time, such as the study of individual persons, language, religion and belief systems, the legal system, literature and art. For Hettner, geography, because it begins with the observation of unitary events (places, landforms, people, grid locations) and then proceeds to generalize about such individual observations, must employ *both* idiographic and nomothetic approaches and it was often difficult to distinguish between geography as a 'science of laws' and geography as a 'science of events' – the chorological and the chronological facets of geography stood by side as equals.[26] This opposition to the neo-Kantian distinction between the *Naturwissenschaften* and the *Kulturwissenschaften* would put Hettner's notion of geography as *Länderkunde* close to Dilthey's definition of the methodology appropriate for the human studies, but it would not thereby indicate that geography, as Hettner saw it, would be considered as *only* a part of the *Geisteswissenschaften*.

Hettner also recognized that the peculiar position of geography, both in terms of its subject matter and its methodology, dictated that observations made in geography would be related to a number of different systems over a number of different time periods. Accordingly he followed his mentors Kirchhoff and H. Wagner in calling it 'natural science with integrated historical elements', and stated that in the study of geography 'man will usually require a more thorough treatment than any one of the others', i.e. the realms of inorganic and organic nature (Hettner, 1927, 14). Statements such as these point out that Hettner at least recognized a historical root for the systematic cultural sciences and that the required blending of chorological and chronological approaches would necessarily involve the geographer in a form of inquiry which was in part historical.[27]

Kraft

Another geographer who was also schooled in philosophy and was a contemporary of Dilthey was Viktor Kraft. In his 1929 essay 'Die Geographie als Wissenschaft',[28] Kraft is obedient to Hettner's view that geography takes the chorological point of view and that it does so with respect to the earth's surface, and also that there are two basic divisions of the subject: general geography (*allgemeine Erdkunde*) which is a law-seeking science and *das Länderkunde*, which, on the contrary, is a 'science of regional individuality' (*regionalen Individualitat*). Here we find some differences from Hettner's definition of *Länderkunde*. For Kraft, *Länderkunde* referred to a science which was 'oriented to singularity' and practised 'like history'.[29] Recognizing that this statement implies a profound dualism between the practice of physical geography as part of general geography and 'historical' geography as part of *Länderkunde*, Kraft states that this dualism is not created by the study of geography but is inherent in the fact that there can be no approach to knowledge intermediary between that of the general and that of the individual. While this is a distinction Dilthey might have made, it is more important to note that Kraft's notion of *Länderkunde* with its 'orientation to the knowing subject' (*und individualisierender Einstellung*) is quite different from that found in Hettner and is very close to what Dilthey called *Erlebnis* (lived experience). Kraft then continues that insofar as the methods of geography are concerned, the methods employed in the study of *Länderkunde* must treat observations not simply as pure sensations but as 'interpreted sensations' and that any type of explanation always concerns individual facts and the particular causes of a complex of phenomena ('um die individuellen Ursachen eines individuellen Tatschenkomplexes', Kraft, 1929, 19). In this type of explanation, there is no factual evidence but rather circumstantial evidence and in this sense, the geographer is much more like a historian in his method of inquiry. Some of Dilthey's ideas are easily recognized here: the fact that to understand any human action one must start with the context of the knowing subject, the idea that lived experience is not simply one act but a whole complex of acts, the notion that all understanding of human experience is achieved in an indirect fashion and the notion that in the act of description, interpretation is already presupposed. Thus it is interesting to note that while Kraft's paper has been cited as evidence of a neo-Humboldtian 'geography as science' view (Schaefer, 1953), it is also apparent that Kraft had quite an enlightened view of other aspects of other aspects of geography and was quite explicit that they strongly resembled history in their methods of inquiry.

The Landschaftskunde school

Well before Kraft's philosophical treatise on the type of knowledge that was available from the study of geography, another major movement in German geography was very much in evidence. The *Landschaftskunde* or 'landscape science' school of geography begun by J. Wimmer (1885), advanced by Otto Schlüter (1906) and propounded by his students (Passarge, Waibel, Bobek, Lautensach and Banse) was a powerful counterpart to the followers of Alfred Hettner. Defining '*landschaft*' as the 'total impact of an area on man's senses' Schlüter chose to focus geography on the interrelations of the features of a region that gave that region its distinctive appearance.[30] Schlüter did not specifically exclude man from his study of *Landschaftskunde* but was clear that the non-material content of an area, the social, economic, racial, psychological and political conditions of an area, were not to be studied as ends in themselves since they were not observable but were to be considered only as secondary factors in an attempt to explain the observable landscape (James, 1972; Dickinson, 1969). Apparently Schlüter also accepted a neo-Kantian distinction between the natural landscape (*Urlandschaft*) and its transformation by human occupation into a cultural landscape (*Kulturlandscaft*). Undoubtedly, Dilthey's views of the study of the *Geisteswissenschaften* as inclusive of the *Kulturlandscaft* would not have been acceptable to Schlüter. However with the maturation of the *Landschaftskunde* school, there were some significant changes. Under the influence of Waibel, the non-material conditions of *landschaft* were alleged as properly the object of study and the term was given a second connotation, that of a 'harmony of related parts' (James, 1972, 231). *Landschaft* was then seen as 'that which stands out from its surroundings (bild) and is important for the creation of man's habitat in a region'.[31] Later, other students of the *Landschaftskunde* school were to admit that 'geography as the study of the transformation of the original natural landscape to the humanized habitat' was a process requiring 'historical interpretation' (Schmieder in Dickinson, 1969, 159) and that although social geography was not to be considered a geography of man, it must concern the origins and influence of 'anthropogenic forces' on social groups as spatial structures (Bobek, 1959, in Dickinson, 1969, 168). Perhaps the most extreme of the *Landschaftskunde* geographers was Banse (1883–1953). Banse claimed that for the most part the 'scientific' regional geographies did not succeed in integrating all the elements of a region since their authors 'suppressed their visual and emotional impressions of the country and the people lest they appear unscientific' (Banse in Fischer *et al.*, 1967, 168). His new geographical doctrine was based not on the rational investigation of the country but on 'inner experience' and was (appropriately) named *Gestaltende Geographie* or creative geography. Calling on geographers to experience the countryside rather than rationally investigate it, he states that the three basic categories of landscape, race and civilization can be understood if geographers would only use their creative faculties to go beyond mere recognition and description and penetrate through the landscape 'to the soul of a country, the indefinable something which hovers invisibly between and above the individual objects and ultimately and truly keeps them together' (Banse in Fischer *et al.*, 1967, 172).

For reasons both academic and non-academic (Banse wrote extensively on the topic of German superiority by world conquest in the mid 1930s), Banse has not been considered important to the mainstream of German geography. Still his emphasis on experiential versus rational investigation is clearly similar to Dilthey's emphasis on the study of *Erlebnis*; the notion of *Gestaltende Geographie* would certainly have been included as part of the *Geisteswissenschaften* and the stress on an interpretative rather than a merely descriptive methodology perhaps best directs his works towards Dilthey's concept of *Verstehen* and the interpretation of texts as basic techniques in the human studies.[32] Dilthey would not have supported Banse's more mystical ideas on how to penetrate 'to the soul of a country' but there is enough similarity of purpose here to suggest that at least some of the practitioners of the *Landschaftskunde* school of geography would be entirely at home with the hermeneutic approach to the study of human action.

Modern geography

With the overwhelming emphasis placed on geography as a spatial science in recent years, it is not surprising that there has been very little serious consideration given to Dilthey as an antecedent to the present form of human geography. There are however, some geographers who are students of the tradition of historical understanding as it applies to geography (Lukermann, Harris, Guelke) and some others whose works exhibit Dilthey's themes in an unmistakable fashion (Lowenthal, Relph, Buttimer, Tuan).

Lukermann has stated quite clearly that 'the geographer's purpose is to understand man's experience; not primarily to judge what is lawful' (Lukermann, 1964, 172). There is for Lukermann a heavy emphasis on the study of *places* as central to geography stressing that 'to emphasize the relative, the cultural, the historical experience of mankind along with the physical attributes of area is to do the complete study of geography' (Lukermann, 1964, 172). Here then we find indirect but very clear reference to the influence of Dilthey; it is the task of geography to 'understand' rather than to 'explain' the phenomena under study and further to understand man's historical experience of place rather than to deal only with the non-human portion of geographic phenomena. There is also the conviction that the study of geography, in order to be complete, must include an historical component or, in Dilthey's terms, we only understand what is present to us because we have already understood similar occurrences in the past. Lukermann has also developed his discussion of neo-Kantian geography in terms of the opposition between the Hettner-Hartshorne school of geography and the historical understanding tradition of Dilthey which attempts to 'relate the development of an event to its experienced context, not as a law, not as a description' but as a result of 'reflective judgement' (Lukermann, 1975, 29). Here we find explicit mention of Dilthey as the proper antecedent of a major school of historical geography, one founded on the notion that it is *Erlebnis* (and not

Erfahrung or *Innerwerden*) that is the proper object of study for human geography.

Perhaps the most extensive treatment of Dilthey's ideas in the geographical literature has been by Cole Harris. He points out that the principal explanations for the development of a particular landscape may not rest in its physical configuration, although this surely is a factor, but rather with society and its values. If this is the case then any 'historian' who wishes to account for a particular landscape must 'place himself in the position of the participants in it by rethinking the thought out of which the event developed' (Harris, 1971, 165). Harris is clear that it is Collingwood and Walsh that he is following here but the influence of Dilthey is explicit: 'the ideas embodied in events compromise a means whereby the interrelations of different events may be considered, a historical account tracing the flow of ideas through many events' (Harris, 1971, 167). It is exactly this sense of 'historical' that Dilthey's tradition speaks of; not the study of past events themselves but the study of the embodiment in those events of certain ideas, values, and beliefs. Thus in Dilthey's *Erlebnis – Ausdruck – Verstehen* formulation, to understand an event one must understand the transmitting tradition embodied or carried through time in that event. Further, Harris states that to understand a 'complex set of relationships bearing on the character of a particular place, is to achieve a synthesizing understanding analogous to that in history' (168). The emphasis on 'synthesizing understanding' is, of course, directly related to Dilthey's notion of *Verstehen* as a principal method of investigating any particular world view or overall perspective on life (*Weltanschauungen*). For Dilthey all of history was a text to be deciphered: the interpreter reduces history to a comprehensive understanding of texts, a type of understanding that expressly seeks out the 'psychic reality' which stands behind the text and of which the text is an expression.[33] Harris' synthesizing understanding seems very similar to this Diltheyian description of the purpose and methods of historical understanding while it does not explicitly elaborate the text-interpretation model of understanding for historical geography.

Of the most recent calls for a humanistic geography, those of Yi-Fu Tuan (1974a, 1976) and Anne Buttimer (1976) can be recognized as based in Dilthey's philosophy of historical understanding.[34] Tuan states that 'humanistic geography' is 'not an earth science' and that with its purpose as the better understanding of man and his condition, humanistic geography properly belongs to the humanities (Tuan, 1976, 266). For Tuan, humanistic geography is 'event-centred'; it is the study of articulated geographical ideas such as 'place', 'territory', 'crowding', 'privacy' and religion and it is particularly concerned with describing the quality of the emotion experienced in specific settings, e.g. the study of place as a centre of meaning for an individual rather than place as described by its geographical coordinates.[35] On this view, humanistic geography would certainly appear to be a part of the *Geisteswissenschaften* and thus would have as its principal focus the study of lived experience as it is related to its context, or *Erlebnis-in-situ*. Just as Dilthey conceived of the task of psychology as seeing the underlying structure in all psychic experience, Tuan, in a similar fashion, conceives of geography as examining 'nuggets of experience' and decomposing them into 'simple themes which can systematically be ordered' thereby giving experience 'an explicit structure' (Tuan, 1976, 274). The humanistic geographer then, needs to be well versed in linguistic usage, 'so he can read, so to speak, between the lines of a text and hear the unsaid in a conversation' (275). Here Tuan indirectly points to hermeneutics as an important principle of humanistic geography. In this he would be confirming Dilthey's conviction that we have a certain implicit understanding of 'psycho-historical reality' and that it is the job of the social scientist to investigate how such 'inner' experiences of 'place', 'home', 'journey', 'territory', etc. become expressed in an outer form and thereby made 'public'.[36] The solution to such inner-outer translation problems was, for Dilthey, the study of exegesis or 'the study and systematic understanding of fixed and relatively permanent expressions of life' (Dilthey, 1900, 232). The more mundane function of a geographer 'to clarify the meaning of concepts, symbols and aspirations as they pertain to space and place' (Tuan, 1976, 275), while not achieving such a complete exegesis, is certainly conceivable as hermeneutic in character.

7 THE NOTION OF 'TEXT'

The importance of the notion of text for Dilthey is based on his dual assumptions that the proper object of understanding in the *Geisteswissenschaften* is the text to be deciphered and understood and that every encounter with a text is likewise an encounter with the transmitting tradition itself. Dilthey thus identified the meaning of a text or action with the subjective intention of its author: to rediscover this subjective intention through the documents, signs, artifacts and life-expressions of the text was the aim of text interpretation in the human studies. Ultimately this understanding of the text for the interpreter carries with it the notion that because historical consciousness is a mode of self-consciousness, that therefore the continuity of tradition embodied in the text will be easily grasped by the interpreter and further that a perfectly adequate comprehension of others will also result, since the text is an expression (a set of external signs) intimately linked to the inner psychic process of the other. Everything in the other is thus understood since everything of the other resembles a text; the study of the historical past and the study of another are thus seen as completely contemporaneous activities (Linge, in Gadamer, 1976b).

While Dilthey's thesis of text-interpretation as a paradigm for the social sciences is perhaps the single most important antecendent to contemporary work in humanistic studies, when followed to its natural conclusion, Dilthey's theory of historical consciousness leads to a position of epistemological relativism. Human nature was at once supposed to be constructed from and known by its external 'life-expressions' embodied in texts. But these life expressions were not the actual historical events themselves, they were only the 'outer side', the words, signs, symbols and gestures, from which we are supposed to infer the 'inner events'. But how are we to know that this inner state of another person in the past is in fact equivalent to our own inner state? Caught in this dilemma of inner versus outer realms, Dilthey ultimately wishes to have an autonomous subject, one who is not affected by his own historicity, an interpreter who can set aside his own prejudices, values and beliefs and simply decipher the text concerned. However, one cannot ignore the intrinsic 'temporarlity' of human existence neither for the actor nor for the spectator and therefore understanding taken as text interpretation has more recently been seen as a text – event involving both the text *and* the interpreter (Gadamer, 1975; Ricoeur, 1977). Dilthey's hermeneutics took the text signs to be indicative of what lay *behind* the signs but modern hermeneutics tends to look *with* a text in terms of what it says to successive generations of interpreters thus involving the text and the interpreter in an on-going process of interpretation. The gist, thrust or force of a text then is a result of a fusion of the historical tradition evident in the text and so interpreted by the reader in his own historical horizon. There is then a sense in which each interpreter returns to the stage of letting the text 'say again' its meanings each time an interpretation is required (Ricoeur, 1977). Hence the meanings of the text are not simply deciphered, they are 'presenced' or mediated through both the past formulations of past interpreters and the present formulations of present interpreters – a fully dialectical process. In the end this process is a productive one in that it often goes beyond the interpretations formerly given to the text and is not reproductive in the sense Dilthey had conceived.

There are several important consequences of this view of text interpretation. First, following Merleau-Ponty, the activity of text-reading involves the interpreter inextricably with the language signs he is reading. Rather than being regarded as objects located outside the interpreter's horizon, linguistic signs, when they are recognized as purveyors of a meaning in a text, become part of the subject (Heelan, 1977). Whenever the signs are read for their meaning in a text they can no longer be regarded as 'neutral', they become infused with the general context of a tradition, or they embody that tradition such that out of a whole gamut of possible meanings, some few meanings are conventionally established.

This active counterpart in the dialectic between text and interpreter is described as follows:

> Before deciphering the language, the signs are outside the subject; after deciphering the language, and to the extent that the language is being used as a text to be read or spoken, the signs are part of the being of the subject. The subject when he reads a text or speaks a language embodies himself in the language signs and his noetic intention operates intentionally through them in order to objectivate a horizon of meaning. (Heelan, 1977, 12)

In contrast to Dilthey's historical understanding where the interpreter stands *outside* the text and attempts to intuit the inner side of the signs in the text, here the 'text lives in its presentations' because its interpretation is 'transsubjective' in Gadamer's terms; it involves the subjectivity of the author, the interpreter and the interpreter-as-subject while reading the text. Taken in this sense understanding a text can be seen as a 'moment in the life of the tradition itself' (Linge in Gadamer, 1976b, 28).

Second, authentic interpretation of the meaning of a text must have something to do with the whole text. Texts are not simply sets of signs, symbols and gestures which are written out and exist apart from the meaning of the text itself; there is an 'idea expressed' or a sense to each text which is disclosed through and in the midst of its parts.[37] Text interpretation requires that this whole be allowed to come forth and be expressed.[38] There is then, not a linear sequence of 'first the parts, then the whole', but rather an entering into the text *via* the parts such that the whole (meaning) is immanent within the parts (Bortoft, 1971, 54). We are thus not 'spectators' of the meaning of a text but active agents in 'a presence which emerges globally so that we find ourselves everywhere within it' (Bortoft, 1971, 61). With this recognition of text interpretation as proceeding 'inside-under' rather than 'outside-over' the text, there is a radical reversal on the part of the interpreter such that we no longer see the text as a set of linguistic signs to which we arbitrarily attach meanings but rather the text *through* its signs as an expression of our intentionality.[39] Just as language itself is never at the focal point of consciousness in the act of speaking rather we use language to speak *about* that which is at the centre of our attention, the meaning of texts as wholes arises not from the symbols themselves but from the attempt to comprehend the symbols within the context of a radical reversal of the subject-object dichotomy. In progressing from the signs as 'what is said' to the whole of the text as 'what the text is about', we begun to encounter *ourselves* as authors of the text-event, an event we call understanding. To continue this self-encounter while being concerned with the understanding of other people's actions is the aim of the text interpretation model for the human studies.

What then for Dilthey was a rather one-sided process, that of compre-
hending symbolized and externalized 'life-expressions' (*Erlebnisausdruck*)
through the activity of *Verstehen* must now be seen as a more complex
process involving the text, the interpreter, his own sense of his immanent
subjectivity and the text as he sees himself as its interpreter. From this
process of text-reading, the 'gist', 'thrust', 'force' or 'direction' of the text
must somehow become evident if the model is to flourish. More specifically, interpretations are built up out of a variety of phenomena attendent
upon the symbols and signs available in the text. Such interpretations
could be expected to contain the welter of information already taken for
granted by the interpreter as part of his biographical situation, his cultural
background, his stock of knowledge, his personal relevance structures
and intentions and most importantly his relationships with other people.
Thus the content of texts is subject to the whole gamut of words, phrases,
symbols, actions, emotions, judgements, memories, theories, desires
and attitudes which the interpreter holds as part of his world, or more
properly as part of his social being in the world. A text then is as fundamental a feature of the everyday world as are the physical and social
objects in it except that a text is an intangible and symbolic feature
whereas the physical and social portions of the world are more easily
indentifiable. In geography, it is also the case that 'texts' are provided by
many other means: any map, questionnaire or equation is likewise a text
for which there is a great deal of agreement as to the specific meaning.[40]
And since the interpretation events required to translate such texts to
other people, or other geographers, take place entirely within a sphere of
signs, a sphere which, like the text itself, has intersubjective origins,
'geographical texts' thus comprise linguistic, social, spatial, temporal and
cognitive dimensions at once. In this sense text interpretation can be seen
as a fundamental way of knowing the geographical world.

HUMAN GEOGRAPHY AS TEXT-INTERPRETATION

The legacy of the Diltheyian tradition of text intepretation is today concerned with the hermeneutical focus on those aspects of linguisticality
that operate in all understanding or 'an unambiguous demonstration of
the continuous process of mediation by which that which is societally
transmitted (the tradition) lives on' (Gadamer, 1976b, 29). To understand
is to understand a text-event in which the past in the form of the text and
the present in the person of the interpreter and his interpretation are
continuously bound up, shot through with each other. There is in this
model of understanding the danger that the interpreter will rather naïvely
call his interpretation of a text the 'correct' one. But there is nothing which
is 'simply there' for the interpreter in a text. There is always a certain
unfinished quality to text interpretation; it is only because the text exists

for the interpreter at all that there can be said to be *any* type of understanding. In a sense, 'everything that is said and is there in the text stands
under anticipations' (Gadamer, 1976b, 121). Thus the opposite of 'what
occurred' or 'took place' for an interpreter reading a text is not what did
not occur or did not happen, but that which has no text, that is 'senseless',
or conveys no message. To objectify a life-expression then, to return to
Dilthey's formulation, thus means to bring it under anticipation, a necessarily circular movement in which we understand what is there *via* the
parts of the texts, but we only see the words, signs, gestures and actions
as individual linguistic expressions. For the human geographer the
problem amounts to this: as an interpreter of lived experience in context
(*Erlebnis-in-situ*), he must fully expect to have to see the text and its
symbols as derivative of a linguistic tradition in which both he and the
author of the text were and are intimately connected, but he must also
expect to have to 'anticipate' or do a considerable amount of 'new seeing'
of senses or possible meanings for the text. 'New seeing' is similar to
reading between the lines or going beyond the author's subjective act of
meaning to a fusion of interpretations taken from the text *and* from the
interpreter's own world.[41] In this type of text interpretation we are likely
to encounter phenomena for which there are no agreed upon definitions
and actions for which there are completely different interpretations. For
example one writer concerned with urban living conditions defines a
'sink' as:

> places of last resort into which powerful groups in society shove,
> dump and pour whatever they do not like or cannot use: auto
> carcasses, garbage, trash and minority groups. (Clay, 1973, 143)

Since the word 'sink' is an extension of a word with other well known
meanings, it does not disclose the experiential qualities of 'being in a
sink' transmitted in this text. The word 'sink' simply does not do justice to
the local junkyard on a rainy afternoon in November or to New York
City's 42nd street on a steamy night in July. To see a sink as a forgotten
bunch of outcasts from society requires a reading of the text which is
event induced; the event being a new interpretation of a previously
known experience, the reference to minority groups, in a completely
different context, i.e. the comprehension of a new type of *Erlebnis-in-situ*.

Some characteristics of text interpretation in human geography

There is by now a growing body of literature in human geography which
gives evidence, directly or indirectly, of its intellectual origins in the
notion of text interpretation. Early work on the actions of people in
landscape situations emanated not from geographers but from the work
of architects and planners (Jackson, 1957; Lynch and Rivkin, 1959; Lynch,

1960; Craik, 1968) and continues today in the tradition of landscape experience (Liebow, 1967; Jackson, 1970; Clay, 1973; Coles, 1971; Newman, 1972) and anthropology (Gladwin, 1970; Lewis, 1972). Geographers have concentrated on the meaning of place and the general linkages between environment, landscape and the mind, (Tuan, 1974a and b, 1976, 1977; Relph, 1976; Appelton, 1975; Lowenthal and Bowden, 1976; Porteous, 1976) and a few pioneering efforts have sought to encounter the geographer as interpreter of his own work (Buttimer, 1972; Michelson, 1973; Ley, 1974; Rowles, 1976; Seamon, 1977). From these works we can note several characteristics of the text interpretation model of understanding:

(a) There is an 'initial credibility' characteristic to the text for the author and for the interpreter which sets off the text from its context – the text is 'noticed' or made apparent in a sense which the background material is not. Liebow's descriptions of the Washington ghettos as a participant-observer or Coles' description of Appalachia leaves no doubt that the observations given are totally credible for the author. As readers of these texts we do not ever have the impression that the experiences related were contrived or 'unreal' in any way. Without this initial credibility, a text has no presence, it is merely an assemblage of signs.

(b) The text exhibits a definable noetic-neomatic structure, that is, the text discloses something about the author and his world as well as what is said about that world. Relph's discussion of 'insidedness', the degree to which the experiencer belongs to and associates himself with a place as a basic feature of identity with place, is an indication of his contact with the tradition of an authentic sense of place found in the work of the French 'géographie humaine' school (Dardel, 1952; Moles, 1971). But Relph's own subjectivity is quite evident in what he identifies as *Kitsch*; the uncritical acceptance of mass values, mediocrity, phoniness and *gemutlichkeit* in general often found as characteristic of specific places. His attack on the 'pornscape of the Barbary Coast' (San Francisco) and the 'ersatz Swiss chalets' of Southern Ontario, quite clearly could be interpreted as his own sense of inauthenticity in the experience of place. Tuan's 'at-homeness' and 'intimate experiences of place' (Tuan, 1977) as well as Bachelard's discussion of 'house', 'nest' and the 'intimate immensity of forests' are similar instances of texts which reveal the author with all his personal values, prejudices and idiosyncracies. To read such texts is to grasp something of the author as well as to see what he is saying about his world.

(c) Texts present both a 'fixed' and an 'open' character, that is the text displays itself conspicuously in some portions and inconspicuously in other portions. This 'relief' of the text is due in part, to what is important in the transmitted tradition evident in the text (a certain 'fixedness' is evident by virtue of the fact that the tradition *is* transmitted), and in part derivative of the emphasis which the author places on certain portions of the text. When we read Tuan's description of Beacon Hill in Boston (Tuan, 1974b, 211), we are not merely trying to reconstruct the world of the writer but rather trying to see how Tuan-as-author of this text sees Beacon Hill. A plausible interpretation gained from a reading of this text would be that the residents of Beacon Hill are very rich, intelligent, proud, inaccessible and come mainly from old, well established families. The 'openness' of the relief in such a text is the fact that simply because Tuan's text gives us this message that new interpretations are not also possible. Were I to meet a Beacon Hill resident this text might provide me with a possible typology with which to start understanding Beacon Hill residents, but it just as easily might not – my interpretation of the text must be kept open, accessible to other ways of seeing the tradition in the text and to other interpretations of the author as author-of-the-text.

(d) Texts, once interpreted, present a stratified or hierarchical structure. There are in our interpretation of texts 'zones of clarity' and relevance; certain portions of the text speak directly to our own thoughts and feelings about the topic of the text while other portions are less clearly related to our interests. Tuan's descriptions of intimate places for Americans as symbolized by the New England church, the Mid-Western town square, the corner drugstore, main street and the village pond (Tuan, 1977, 147) do not disguise the availability of countless other similar symbols more relevent to readers in Europe, South America or the Far East. Through its interpretation the text takes on a relevance structure such that the reader is brought to familiarity with certain portions of the text more than with other portions. The tradition embodied in the text is thus emergent in pieces and yet at the same time it is guided by the sense of wholeness or direction evident in the text. Working through this part-whole paradox and seeing the 'contours' of the text is what the interpretation event is about – the dialectics of text interpretation.

(e) Text interpretation is a type of understanding which we already possess as part of our own being in the world; it is part of what we do to 'operate' our lives. The tradition begun by Dilthey and taken up by modern hermeneutic philosophers such as Gadamer and Ricoeur is precise in its statements that *Verstehen* is not simply a method of 'scientific' inquiry (although it is that as well), but it is the most common of our linguistic experiences and that which goes on continuously in everyday life.[42] The understanding of themes such as 'place', 'at-homeness', 'territory', and 'insidedness' is carried out in the context of the everyday

E

world – there are no 'special ways' in which the geographer understands these themes. Hence it is the understanding of *Erlebnis-in-situ* which is the object of humanistic geography and it is the understanding of the reality of this object precisely within language and within texts in particular that is the work of the humanistic geographer.

Within this view of humanistic geography as text interpretation there still exists the dilemma that Dilthey faced. If we accept the premises that: (1) human geography takes place entirely within a sphere of signs, (2) that some of these signs are found together in texts whether spoken, written, gestured or acted out, and (3) it is the job of the human geographer to interpret such texts as a spectator in order to make certain statements about actors operating within the texts, and further (4) to communicate the meanings of such phenomena as he should deem important back to the actors involved, then we cannot escape the conclusion that the assumption of a sign system in which perception, speaking, interpretation and reporting take place has as much bearing on the reported phenomena under study as the phenomena themselves. Thus perception, and hence interpretation, when considered 'in' a reflexive symbolic system of texts cannot easily be divorced from their products. What is perceived then is not a text or a sign or a symbol 'detached' from or pointing toward a specific meaning, rather the text and the actions, thoughts, and emotions it embodies and the interpretation it is given are often quite inseparable – the 'doing' of the action, of 'saying' of the text and the 'knowing' of the meaning are all very much overlapped. Hence what is required of geographers who take text interpretation as defining what they 'do' when they operate as geographers is not to obtain directly the meanings to which the texts refer but rather the attempt to rediscover those meanings from the text, from the transmitted tradition in the text and from their own worlds.

Notes

[1] This is not an entirely accurate statement. The work of the German life philosophers has had a profound influence on both the philosophy of history and modern idealism. In this connection their ideas are not unknown to many historical geographers, e.g. Harris's (1971) notion of 'historical synthesis' is a derivative of the works of Croce, Collingwood and Walsh, and Guelke's (1974) 'idealist alternative' for human geography is based directly on Collingwood (see part 6).

[2] Among the earlier works are: *The life of Schleiermacher* (1867–70), *An introduction to the human studies* (1883), and the *Conception and analysis of human nature in the 15th and 16th centuries* (1891–3). Dilthey's pre-hermeneutic phase was marked by works such as 'The Rise of Hermeneutics' (1900) in *New Literary History: A Journal of Theory and Interpretation* 3, no.2 (1972), 229–44, *Das Erlebnis und die Dichtung* (1906) and *Der Aufbau der geschichtlichen Welt in den Geisteswissenschaften* (1910).

3 *Erlebnis*, often translated as 'lived' or 'felt' experience was defined in various ways in Dilthey's works. Since Dilthey regarded *Erlebnis* as central to the understanding of any poetic imagination, the term refers in his early works to what is felt in experience. But Dilthey is clear that *Erlebnis* refers to totally conscious activities, particularly those which form the connectedness of psychic life. He regards it as a 'unit of experience which is immediately recognizable as manifesting a meaningful relation to human life' (*Gesammelte Schriften* VI, 314). One of its characteristics is that *Erlebnis* constitutes an intersection of all the powers of consciousness and that *Erlebnis* is always given in certainty in contrast to some of our perceptions of the outer world (Makkreel, 1975, 147). Thus *Erlebnis* is not merely to be equated with 'inner experience' but is rather 'inner experience in terms of its natural relations to outer experience' (Makkreel, 1975, 283).

4 References to the collected papers of Dilthey will be referred to using the abbreviation *G.S.* Volumes I–XII were published by B. G. Tuebner, Stuttgart and by Vandenhoeck and Ruprecht, Gottingen 1922–36. Volumes XIII–XVII were published by Vandenhoeck and Ruprecht, Gottingen 1966–74. Selected passages also appear in H. A. Hodges, *Wilhelm Dilthey: An introduction* (New York, 1969), 109–56.

5 For example, Dilthey writes: 'If there were something timeless behind life, that runs through a past, present and future, it would be an antecedent of life. Therefore, it would be the condition of the total structure of life. This antecedent would be just what we do not experience and consequently, a realm of shadows.' *G. S.* V, 5).

6 The term 'Geisteswissenschaften' has no satisfactory English equivalent. It has been used as a translation for J. S. Mill's term 'the moral sciences', but its origin dates to the historian J. G. Droysen in the *Geschichte des Hellenismus* (1843). A literal rendering of the term would be 'science of the spirit' (Geist) but it was also used by certain scholars, e.g. Krause, to mean 'science of reason' and for most Hegelians, it would mean 'philosophy of spirit' (Makkreel, 1975, 36).

7 Dilthey states: 'By life-expressions, I understand not only those expressions which signify or mean something in particular; I specifically include those which, without so intending, make mental life comprehensible' (*G.S.* VII, as quoted in Gardiner, 1959, 213). These 'life-expressions' are directly given in experience according to Dilthey and thus no part of them is a product of our hypothesizing activity, an activity which presupposes that the elements of the hypotheses be not only limited to a certain number but also defined in a certain specific manner. In this sense the elements of the natural sciences (*Naturwissenschaften*) are given only through inference whereas those in the *Geisteswissenschaften* are given as primary and fundamental (*G.S.* V, 143–4).

8 In his earlier work, Dilthey was not exactly certain as to what was to be included in the human studies, e.g.

> Facts, theorems, value-judgements and rules: of these three classes of propositions the human studies are composed. And in the relation between the historical, the abstract theoretical and the practical tendencies of the conception of the human studies is a common basic relationship pervading them all. (*G. S.* I, 26)

But by 1910, Dilthey had a clear conception of the differences between the natural sciences and the human studies:

> We can now mark off the human studies from the natural sciences by quite clear criteria. These lie in the attitude of mind described above, by which,

in contrast with natural scientific knowledge, the object of the human
studies is constituted. Mankind, if apprehended only by perception and
perceptual knowledge, would be for us a physical fact and as such it would
be accessible only to natural-scientific knowledge. It becomes an object for
the human studies only in so far as the human states are consciously lived,
in so far as they find expression in living utterances and insofar as these
experiences are understood. (*G. S.* VIII, 86–7).

[9] This is especially pronounced in Dilthey's exchanges with the neo-Kantians over
the nature of the human studies. The *Kulturwissenschaften* or cultural sciences
were, on Rickert's account, 'individuating' sciences and did not include
history. History, Rickert asserts, does employ some universal concepts to select
out what is essential from what is irrelevant and universal concepts can be
combined to create individual complexes (such as the study of Napoleon); these
complexes in history are known as values. See Makkreel, 1969; Hodges, 1944,
chapter 5.

[10] Makkreel's interpretation of Dilthey's use of the term 'transcendental' is that it is
here used in a more general Kantian sense to mean how the consideration of
any inner experience naturally leads to reflectivity and thus would be derivative
of Kant's use of the term in the *Critique of Judgement* rather than as Kant's use of
the term to mean pre-condition of experience itself (Makkreel, 1975, 223).

[11] The term *Erlebnis* is coined from the verb *erleben* meaning 'to experience,
particularly in individual instances'. The more general term *Erfahrung*, meaning 'experience in general' was used in German before Dilthey's introduction of
the more specific term *Erlebnis*. It is thought that Dilthey's use of *Erlebnis* was
derivative of Goethe's term *Erlebnisse* (Palmer, 1969, 107).

[12] Hodge's interpretation of this sequence of components within the unit of action
is as follows: 'I cannot recognize a thing without being interested in it, and
having feelings and/or desires about it. I cannot feel unless I have an idea of the
object and feeling tends to pass over into action. I cannot act unless I know the
situation and my own aim and action is usually motivated and always accompanied by feeling' (Hodges, 1944, 43). Some of the difficulty inherent in
Dilthey's task of writing a 'Critique of Historical Reason' is evident in this
passage. Having committed himself to the idea that *Erlebnis* is a 'reality of
what-is-there-for-me before experience becomes objective', Dilthey then proceeds to destroy this prior unity by postulating the three temporally sequential
categories of experience and state that feelings for an experience must somehow depend on our 'objectifying' the cognition of an experience first. But the
experience (such as being mugged) is what 'lifts out' various events and
feelings from one's personal flow of experiences and gives them a certain
coherence. Dilthey's attempted 'objectification' of lived experience may have
been misguided by comparison to modern hermeneutics but his central insight
lay in seeing experience as a realm in which the world and our experiences in
that world are, at their origin, inseparable.

[13] The similarity here between Dilthey's concept of *Erlebnis* and Husserl's use of the
same notion is extensive. Both Dilthey and Husserl agreed that *Erlebnis*, not
Erfahrung or *Innewerden* (inner perception) was the proper object of study for
the human studies. But while Dilthey's notion had a natural relation to objects
and experiences in the 'outer world', Husserl's construct was intuitively evident in the *épochè*, it was purely given (Makkreel, 1975, 283).

[14] This use of 'intentionality' could be seen to anticipate Husserl's use of the term.

Husserl's 'intentional analysis' referred to the method by which cognitive acts were used as starting points to go back to their original correlates in lived experiences. In this analysis, intentionality is the characteristic property of the consciousness, always directing the consciousness toward some object and always codetermining the character and meaning of the act in which that object appears. The purpose of this analysis was to obtain access to 'the world as it manifests itself in primordial experience, . . . the "natural" world, the world of immediate experience (Lebenswelt)' (Kockelmans, 1967, 34).

[15] Dilthey writes: 'When we look back in memory, we apprehend the nexus of elapsed phases of the course of life through the category of meaning' (*G. S.* VII, 201). While it is true that Dilthey held that the past is the most meaningful dimension of time and that hence the process of understanding must involve some regressive movement, this is so primarily because Dilthey saw (correctly) that the past is the only mode of time that can be concretely apprehended as a whole made up of several related parts. (Makkreel, 1975, 383). But it can be argued that present actions also hold meaning and that any plan for the future is also meaningful and hence that understanding can be thought of as a forward-moving event as well as a regressive movement.

[16] Or as Makkreel states: 'Self-understanding is thus not directly fed into *Erlebnis*, but requires that I regard myself as a text to be interpreted' (Makkreel, 1975, 256).

[17] The following passages describe the operation of *Verstehen*:

Understanding is not a mere act of thought but a transposition and re-experiencing of the world as another person meets it in lived exerience. It is not a conscious, reflexive act of comparison but an operation of silent thought which accomplishes a prereflexive transposition of oneself into the other person (*G. S.* VII, 191).

In short it is through the process of understanding that life in its depths is made clear to itself and on the other hand we understand ourselves and others only when we transfer our own lived experience into every kind of expression of our own and other people's life (*G. S.* VII, 86–87).

[18] The translation of *Verstehen* as 'emphatic understanding' is generally attributed to H. P. Rickman in *Pattern and Meaning in History* (New York, 1962), and it is this translation which has been used in certain critiques of the notion, notably by Nagel (1953). Makkreel, however, points out that both Nagel's and Abel's (1948) rendering of *Verstehen* as 'understanding an action . . . if we can apply to it a generalization based upon personal experience' are incorrect. Dilthey's operation of *Verstehen* attempted to draw out the meanings of certain actions by placing them in contexts where the actions could be examined indirectly and reflectively. (Makkreel, 1975, 252n).

[19] Dilthey writes:

. . . we understand individuals by virtue of their affinitites with one another, the common factors which they share. This process presupposes the relation between human nature in general and individuality which, on the basis of the general, branches out into the multiplicity of mental-existences. The problem which we continually solve in practice when we understand is that of inwardly living through as it were this movement towards individuality. The material for the solution of this problem comprises the particular data as induction brings them together. Each is something individual and is grasped as such. It therefore contains an element which makes it possible to seize the determinate individuality of the whole' (*G. S.* VII, 212).

[20] Friedrich Schleiermacher, *Hermeneutik*, ed. H. Kimmerle (Heidelburg, 1959); Friedrich Schleiermacher, *Hermeneutik und Kritik*, ed. F. Lucke (Berlin, 1938).

[21] Husserl states: 'The formation of historical consciousness destroys . . . a belief in the universal validity of any of the philosophies that have undertaken to express in a compelling manner the coherence of the world by an ensemble of concepts' (Lauer, 1965, 124).

[22] And further, 'Insofar as an object is felt (erlebt) it is apprehended as it exists for me and its relations to my existence are what counts and these references to it are real regardless of whether it can be confirmed to stand there over against me in the natural world' (*G. S.* VII, 27, 33).

[23] 'Meaningfulness fundamentally grows out of the relation of part and whole that is grounded in the nature of lived experience' (*G.S.* VIII, 233).

[24] *Das Erlebnis und die Dichtung* (B. G. Teubner, Stuttgart, 1960) and *Der Aufbau der geschichtichen Welt in den Geisteswissenschaften* (B. G. Teubner, Stuttgart, 1927) first published in 1910.

[25] Dilthey was quick to point out that these dynamic systems were not the same as those found in nature:
A dynamic system (Wirkungszusammenhang) differs from the causal system (Kausalzusammenhang) of nature in that it produces values and realizes ends according to the structure of psychic life. . . This I call the immanent teleological character of spiritual dynamic systems, by which is meant the nexus of functions grounded in their structure (G. S. VII, 153).

[26] Compare Hettner: 'This combined observation of nature and mankind is not 'dualistic' as the combination of nature and man in the general science of the earth (Erdwissenschaft). One can correctly speak of dualism only when the combination of different things in one science carries a difference of opinion and discord into it. The country-oriented (länderkundlich) or chorological concept of nature and man is, however . . . alike in all esential points and does not lead to two different parts of geography at all' (Hettner, 1927, 14).

[27] Compare Hettner: 'The present can always be understood from the past' and 'Geography needs a genetic (genetischer) concept but it must not become history' (Hettner, 1927, 18) with Dilthey: 'The past is always a permanently enduring present for us' (Rickman, 1976, 221), and 'To grasp the meaning of a moment in the past, we see it through the present' (Rickman, 1961, 106).

[28] *Enzyklopädie der Erdkunde*, O. Kende (Leipzig) 1929, 1–22

[29] Kraft: 'in der Länderkunde dagegen als eine Individualwissenschaft, eine Wissenschaft, die sich, auf Individuelles als solches richtet wie die Geschichte' (Kraft, 1929, 12)

[30] Schlüter: 'One should limit human geography to the form and arrangement of earth-bound phenomena as far as they are perceptible to the senses' (Schlüter in Dickinson, 1969, 129).

[31] It is this emphasis on *habitat* and *bild* which was transported to the US school of landscape geography and is seen principally in the works of Carl Sauer and J. K. Wright.

[32] According to Banse, 'the final tool is the word which describes and explains, imitates, condenses and interprets in order to warrant as impressive and vivid a presentation as possible' (Banse in Fischer, 1967, 173).

[33] Compare Dilthey: *Verstehen* is 'that process by which we intuit behind the sign

given to our senses, that psychic reality of which it is an expression' (Dilthey, 1900, 232) and also Makkreel: 'Self-understanding . . . requires that I regard myself as a text to be interpreted' (Makkreel, 1975, 256).

[34] When geographers use the term 'humanistic' or the notion of the study of space from the humanistic perspective, they have sometimes referred to its more general meaning as that study of spatial experience whch makes use of empathy, intuition, feelings, emotions, desires and other non-empirical methods of study (Entrikin, 1976). In this they would be in accord with other disciplines such as sociology and psychology where phenomenology and existential phenomenology have provided the major thrust for the humanist movement (Berger, 1963; Bruyn, 1966; Maslow, 1968; May, 1960; Wann, 1964; Gendlin, 1962). At other times the term 'humanistic' has often been used to denote any reactive account of spatial experience which does not emanate from the positivistic methodology of geography practised as a science. While this use of the term would include the work of such humanistic geographers as Yi-Fu Tuan, Relph and Buttimer, it would exclude that of Sauer, Lowenthal, Spate, van Paassen, and Hägerstrand. Such problems with the use of the term 'humanistic' when applied to geography indicate that no clear definition of the term is apparent in the discipline as a whole.

[35] Compare Tuan: 'The material includes the nature and range of human experience and thought, the quality and intensity of an emotion, the ambivalence and ambiguity of values and attitudes, the nature and power of the symbol and the character of human events, intentions and aspirations' (Tuan, 1976, 274).

[36] Dilthey's term for this 'making public' of inner experiences was *Nacherleben* or 're-experiencing' meaning to find the inner meaning of an external objectification. Ideally this re-experiencing is an outgrowth of the complete *Verstendun* of an objectification such that in the end, an interpreter can understand the author better than he understood himself (Dilthey, 1900, 244; Makkreel, 1975, 328).

[37] Compare Makkreel: 'The primary framework for understanding troublesome propositions is the entire text' (1975, 271).

[38] James Edie uses an interesting metaphor to describe this activity; he speaks of the meaning of the text as being 'pressed out' of the text by its reading (Edie, 1976, 158).

[39] 'We are accustomed to imagining that we are outside of language which stands over against us as an object of awareness. Consequently, we cannot notice that this position is counterfeit, because we cannot notice that the "I" which seems to itself to see, know and manipulate language is itself a linguistic structure, a condensation in language' (Bortoft, 1971, 67).

[40] For example,
> Now maps are often as beautiful as they are useful, but even the best of them is merely a symbolic representation of selected facts. They are built up of layered distributions which are necessarily incomplete. A map leaves out much more than it can ever contain and a great deal has to be read into it before it can mean very much. In short, it is an auxiliary, not a fundamental element
(Gauld, 1941, 546).

[41] L. W. Beck describes a similar stance taken by the 'critic' wherein he attempts not only to see the actor's or agent's perspective in an action but also to see that of a spectator (Beck, 1965).

[42] In Gadamer's terms this is very much like a game: '. . . it (language) is by itself

the game of interpretation that we are engaged in everyday. In this game
nobody is above and before all the others; everybody is at the centre, is "it" in
this game. Thus it is always his turn to be interpreting' (Gadamer, 1976b, 32).

References

Abel, T. 1948: The operation called verstehen. *American Journal of Sciology* 54,
    211–18.
Anscombe, G. E. M. 1957: *Intention* (Oxford).
Appleton, J. 1975: *The experience of landscape* (New York).
Bachelard, G. 1969: *The poetics of space* (Boston).
Beck, L. W. 1965: Agent, actor, spectator and critic. *The Monist* 49, 167–182.
Berger, P. 1963: *Invitation to sociology: a humanist perspective* (New York).
Bobek, H. and Schmithüsen, J. 1949: Die Landschaft in logischen System der
    Geographie. *Erdkunde* 3, 112–120.
Bollnow, O. 1958: *Die Lebensphilosophie* (Berlin).
Bortoft, H. 1971: The whole: counterfeit and authentic. *Semantics* 9, no. 2, 43–73.
Bruyn, S. 1966: *The human perspective in sociology* (Englewood Cliffs, New Jersey).
Bunge, W. 1966: *Theoretical geography. Lund Studies in Geography, Series C*, 1 (Lund).
Buttimer, A. 1972: Social space and the planning of residential areas. *Environment
    and Behaviour* 4, 279–318.
Buttimer, A. 1976: Grasping the dynamism of life-world. *Annals of the Association of
    American Geographers* 66, no. 2, 277–92.
Chorley, R. J. and Haggett, P. 1967: *Models in geography* (London).
Clay, G. 1973: *Close-up: how to read the American city* (New York).
Coles, R. 1967: *Migrants, sharecroppers, mountaineers* (Boston).
Coles, R. 1971: *The south goes north* (Boston).
Craik, K. 1968: The comprehension of the everyday physical environment. *Journal
    of American Institute of Planners* 34, no. 1, 29–37.
Croce, B. 1921: *History, its theory and practice.* Trans. D. Ainslie (New York).
Dardel, E. 1952: *L'Homme et la Terre* (Paris).
Dickinson, R. 1969: *The makers of modern geography* (London).
Dilthey, W. 1914–74: *Gesammelte Schriften* 17 vols. See note 4.
Dilthey, W. 1883: *Einleitung in die Geisteswissenschaften: Versuch einer Grundlegung
    für das Studium der Gesellschaft und der Geschichte.* B. Groethuysen, editor, fourth
    edition 1959.
Dilthey, W. 1900: The rise of hermeneutics. Trans. F. Jameson, in *New Literary
    History* 3, no. 2 (1972), 229–44.
Dilthey, W. 1906: *Das Erlebnis und die Dichtung: Lessing, Goethe, Novalis, Holderlin,*
    13th ed., Stuttgart and Gottingen, 1957.
Dilthey, W. 1910: *Der Aufbau der Geschichtlichen Welt in den Geisteswissenschaften.* B.
    Groethuysen, editor, second edition (Gottingen) 1958.
Downs, R. 1970: Geographic space perception: past approaches and future
    prospects. *Progress in Geography* 2 (London) 65–108.
Edie, J. 1976: *Speaking and meaning: the phenomenology of language* (Bloomington,
    Indiana).
Entrikin, J. 1976: Contemporary humanism in geography. *Annals of the Association
    of American Geographers* 66, 615–32.
Fischer, E., Campbell, R. and Miller, E. 1967: *A question of place* (Arlington,
    Virginia).
Gadamer, H. G. 1975: Hermaneutics and Social Science. *Cultural Hermeneutics* 2,
    307–36.

Gadamer, H. G. 1976a: The problem of historical consciousness. Trans. J. L. Close, *Graduate Faculty Philosophy Journal* 5, 1–52.
Gadamer, H. G. 1976b: *Philosophical hermeneutics.* Trans. D. E. Linge (Los Angeles).
Gardiner, P. ed. 1959. *Theories of history* (New York).
Gauld, T. 1941: Towards a new geography. *Nature* 147, 546–8.
Gendlin, E. 1962: *Experiencing and the creation of meaning* (New York).
Glacken, T. 1967: *Traces on the Rhodian shore* (Berkeley).
Gladwin, T. 1970: *East is a big bird: navigation and logic on Puluwat Atoll.* (Cambridge, Mass.).
Guelke, L. 1974: The idealist alternative in human geography. *Annals of the Association of American Geographers* 64, no. 2, 193–202.
Haggett, P. 1965: *Locational analysis in human geography* (London).
Harris, C. 1971: Theory and synthesis in historical geography. *Canadian Geographer* 15, no. 3, 157–72.
Hartshorne, R. 1939: *The nature of geography.* (Lancaster Pa)
Hartshorne, R. 1955: Exceptionalism in geography re-examined. *Annals of the Association of American Geographers* 45, no. 3, 205–44.
Harvey, D. 1969: *Explanation in geography* (London).
Heelan, D. 1977: Hermeneutics of experimental science in the context of the life-world. In Ihde, D. and Zaner, R. M. *Interdisciplinary phenomenology* (The Hague) 7–50.
Hettner, A. 1927: *Die Geographie: ihre Geschichte, ihr Wesen und ihre Methoden.* English translation: Geography, its history, its nature and its methods, L. Seidler, 1927 (Breslau).
Hettner, A. 1929: Unsere Auffassung von der Geographie. *Geographische Zeitschrift* 35, 486–91.
Hodges, H. 1944: *Wilhelm Dilthey: an introduction* (New York).
Hodges, H. 1952: *The philosophy of Wilhelm Dilthey* (London).
Holborn, H. 1950: Wilhelm Dilthey and the Critique of Historical Reason. *Journal of the History of Ideas* 11, 93–118.
Husserl, E. 1911: Philosophy as rigorous science. In Husserl, E. *Phenomenology and the crisis of philosophy*, Trans. Q. Lauer (New York) 71–147.
Jackson, J. B. 1957: The stranger's path. *Landscape* 7, no. 1, 29–35.
Jackson, J. B. 1970: *Landscapes: selected writings*, E. Zube, editor. (Cambridge, Mass.).
James, P. 1972: *All possible worlds* (Indianapolis).
Kockelmans, J. J. 1967: *Phenomenology: The philosophy of Edmund Husserl and its interpretation* (New York).
Kraft, V. 1929: Die Geographie als Wissenschaft. In Kunde, O., editor, *Enzyklopädie der Erdkunde* (Leipzig), 1–22.
Lauer, Q. 1965: *Phenomenology: its genesis and prospect* (New York).
Lewis, D. 1972: *We, the navigators* (Honolulu).
Ley, D. 1974: *The Black inner city as frontier outpost: images and behaviour of a Philadelphia neighborhood* (Washington: Association of American Geographers, monograph no. 7)
Liebow, E. 1967: *Tally's Corner* (Boston).
Lowenthal, D. and Bowden. M. 1976: *Geographies of the mind* (Toronto).
Lukermann, F. 1964: Geography as a formal intellectual discipline and the way in which it contributes to human knowledge. *Canadian Geographer* 8, no. 4, 167–72.

Lukermann, F. 1975: *The history and philosophy of science*. Paper presented at the Association of American Geographers Conference, Milwaukee, Wisconsin.

Lynch, K. 1960: *The image of the city* (Cambridge, Mass.).

Lynch, K. and Rivkin, M. 1959: A walk around the block. *Landscape* 8, no. 3, 24–34.

Makkreel, R. 1969: Wilhelm Dilthey and the neo-Kantians: the distinction of the Geisteswissenschaften and the Kulturwissenschaften. *Journal of the History of Philosophy* 7, no. 4, 423–40.

Makkreel, R. 1975: *Dilthey: philosopher of the human studies* (Princeton).

Maslow, A. 1968: *Towards a psychology of being*. (New York).

May, R. 1960: *Existential psychology* (New York).

Michelson, W. 1973: Discretionary and non-discretionary aspects of activity and social contract in residential selection. In Bourne, L. S. 1973: *The form of cities in central Canada* (Toronto), 180–98.

Moles, A. 1971: *Le Kitsch* (Paris).

Nagel, E. 1953: On the method of Verstehen as the sole method of philosophy. *Journal of Philosophy* 50, no. 5, 154–7.

Newman, O. 1972: *Defensible space* (New York).

Palmer, R. 1969: *Hermeneutics* (Chicago).

Porteous, J. 1976: Home: the territorial core. *Geographical Review* 66, no. 4, 383–90.

Relph, E. 1970: An inquiry into the relations between phenomenology and geography. *Canadian Geographer* 14, no. 3, 193–201.

Relph, E. 1976: *Place and placelessness* (London).

Rickman, H. P. 1961: *Meaning in history: W. Dilthey's thoughts on history and society*. (London).

Rickman, H. P. 1976: *Dilthey: selected writings* (New York).

Ricoeur, P. 1973: The model of the text: meaningful action considered as text. *New Literary History* 5, no. 1, 91–117.

Ricoeur, P. 1977: *Interpretation theory: discourse and the surplus of meaning* (Fort Worth, Texas).

Rowles, G. 1976: *Exploring the geographical experience of older people*. Ph.D. dissertation, Clark University, Worcester, Mass.

Ryle, G. 1949: *The concept of mind* (London).

Saarinen, T. 1969: *Perception of environment*. Commission on College Geography, resource paper no. 5. (Washington).

Samuels, M. 1971: *Science and geography: an existential appraisal*. Ph. D. dissertation, University of Washington, Seattle.

Schaefer, F. 1953: Exceptionalism in geography. *Annals of the Association of American Geographers* 43, 226–49.

Schlüter, O. 1920: Die Erdkunde in ihrem Verhältnis zuden Natur- und Geisteswissenschaften. *Geographische Anzeiger* 21, 145–52.

Seamon. D. 1977: Movement, rest and encounter: a phenomenology of everyday environmental experience. Ph. D. dissertation, Clark University, Worcester, Mass.

Tatham, G. 1951a: Geography in the nineteenth century. In Taylor, G., 1951: *Geography in the twentieth century* (London).

Tatham, G. 1951b: Environmentalism and possibilism. In Taylor, G., 1951: *Geography in the twentieth century* (London).

Taylor, G. 1951: *Geography in the twentieth century* (London).

Troeltsch, E. 1922: *Der Historismus und seine Probleme*. (Tubingen).

Tuan, Yi-Fu 1974a: Space and place: humanistic perspective. *Progress in Geography* 6, 211–252.
Tuan, Yi-Fu 1947b: *Topophilia* (Englewood Cliffs, N. J.).
Tuan, Yi-Fu 1976: 'Humanistic Geography'. *Annals of Association of American Geographers* 66, 266–76
Tuan, Yi-Fu 1977: *Space and place: the perspective of experience* (Minneapolis).
Tuttle, H. N. 1969: *Wilhelm Dilthey's philosophy of historical understanding.* (Lieden).
van Valkenburg, S. 1951: 'The German school of geography' In Taylor, G., 1951: *Geography in the twentieth century* (London).
Wann, T. Ed. 1964: *Behaviourism and phenomenology* (Chicago).

8

Peter Kropotkin, the Anarchist Geographer

Myrna Margulies Breitbart

INTRODUCTION

In 1861, Peter Kropotkin stunned his aristocratic Russian father and classmates by choosing to serve a military tour as an officer in Siberia. This decision was motivated by an intense curiosity about the physical geography of the area and by a desire to improve the conditions of prisoners and reform town government. Kropotkin's experiences in Siberia were to have a profound effect on his life.

While in Siberia, Kropotkin was impressed by the spirit of equality and self-sufficiency of Russian peasants (1899, 168). He was disheartened, however, by the negative effects which political centralization had on spontaneous local expression, and by the failure of administrative reform to produce any real change for peasants or political exiles. At the age of twenty-two, therefore, he redirected his energy towards an exploration of the physical environment. Subsequent travels and discoveries established his reputation amongst professional geographers worldwide. Expeditions throughout northern Asia, the Arctic seas, and Finland enabled him to develop credible and original theories of glaciation, desiccation, and orography. While on these travels, he also found pleasure in the harmonies of the natural environment and the apparent oneness of human beings with nature (1899, 116). Through these experiences, he developed an ambition to write a monumental physical geography of northern Europe, and to immerse himself in 'pure' science.

Despite a great enthusiasm for his work, Kropotkin soon discovered that he could not conduct his explorations in a social vacuum – that he could not observe the physical landscape without also viewing the despicable social and economic conditions of Russian peasants and workers. He began to question his right to experience the joy of sorting and analysing objective facts without confronting questions of social justice:

> Science is an excellent thing. I knew its joys and valued them – perhaps more than many of my colleagues did. Even now as I was looking on the lakes and the hillocks of Finland, new and beautiful generalizations arose before my eyes . . . a grand picture was rising and I wanted to draw it, with the thousands of details I saw in it; . . . to open new horizons for geology and physical geography.
>
> But what right had I to these highest joys when all around me was nothing but misery and struggle for a mouldy bit of bread; . . . all these sonorous phrases about making mankind progress, while at the same time the progress-makers stand aloof from those whom they pretend to push onwards, are mere sophisms made up by minds anxious to shake off a fretting contradiction. (1899, 237–41)

Because of this growing awareness, he rejected in 1871 a once eagerly-awaited offer by the Russian Geographical Society to become its secretary (1899, 235, 239–41). Instead, he returned to St. Petersburg to become active in anti-Czarist revolutionary activities. A sensitive pursuit of geographical interests in Siberia thus led Kropotkin to perceive the need for revolutionary change in society.

Many have identified the Siberian period as a turning point in Kropotkin's life. Some have also suggested that revolutionary interests were to mark an end to Kropotkin's brilliant career 'as an original geographer':

> Henceforward he would undertake no journeys or exploration . . . Later . . . he wrote . . . many articles in English and foreign scientific magazines and delivered lectures . . . But he never expanded his knowledge in an original way, or used it for the production of further geographical theories. (Woodcock and Avakumović, 1950, 90)

This interpretation results from a limited appreciation of the potential scope of the discipline of geography. Kropotkin's most important and original contributions to geography – his theories of mutual aid, human ecology, and decentralization – came *after* he escaped from prison in Russia and went into revolutionary exile. From 1874 on, a sense of justice and a desire for free cooperative expression guided the direction of his scientific work. Studies of those environments which support economic and social cooperation also provided a foundation for the later development of his anarchist political beliefs.

Thus Kropotkin did not abandon his geography to become an anarchist in 1871. Rather, he set out to expand the range of geographic inquiry to encompass social critique. From then on, he devoted his life to two revolutions: one in social and economic relations, the other in the discipline of geography itself.

KROPOTKIN AS SOCIAL ANARCHIST

Given the personal and comprehensive focus of social anarchism, it is significant that Kropotkin's introduction to it came, not through the literature, but from a brief personal encounter with a community of anarchist watchmakers in the Jura Mountains of Switzerland in 1872. Their independence of thought, capacity for disentangling complex social questions, and commitment to a cooperative revolutionary way of life convinced Kropotkin to adopt an anarchist version of socialism (1899, 286–7).

Social anarchists maintain a firm belief in the capacity of people to organize their lives without structures of domination and subordination – to coordinate everything from a family to an economy on a cooperative participatory basis. Most socialists focus on the need to replace repressive economic structures with common ownership of the means of production and distribution according to need. For anarchists, freedom is the relationship of people to a total life process, and the elimination of authoritarian relationships wherever they arise – in the workplace, in the home, in the school, or in social situations.

The moral principle in nature

Kropotkin's social anarchism, or anarcho-communism, seeks to demonstrate the cooperative basis of human nature when set within an environment remote from centralist and authoritarian influence. It also emphasizes the importance of freedom and mutual aid as progressive forces in human evolution. Unlike individualist anarchists who stressed the incompatibility of personal freedom with social responsibility (Stirner, 1845), Kropotkin recognized the importance of basing personal autonomy on a strong communal foundation. Enrichment and the growth of the human personality depend upon an identification with the interests of a larger group, and the cultivation of an ever-expanding multiplicity of social relationships.

Thus, Kropotkin believed that true individualism can only be cultivated by the conscious and reflective interaction of people with a social environment which supports their personal freedom and growth (Kropotkin, 1924, 25; Hoffman, 1972, 169). Real freedom does not mean having power over others and true equality does not require the stifling of difference. Rather, there is a need for *unity in diversity* – a sense of mutual dependence on others for collective action, but also an opportunity to express individual difference.

Kropotkin was encouraged to develop his concept of 'mutual aid' by two related philosophical developments in the 1890s – the rise of Social Darwinism which acclaimed the superiority of competition over cooperation in human relationships, and arguments by philosophers, the most extreme example being Nietzsche, which seemed to deny the existence of morality altogether. Kropotkin describes evolutionary history in *Mutual aid* (1914) as a struggle between cooperative and competitive forces. The dominant interpretation of Darwin's *Origin of species* (1859) at that time considered Nature as,

> An immense battlefield upon which one sees nothing but an incessant struggle for life and an extermination of the weak . . . by the strong. (Kropotkin, 1924, 12)

Kropotkin, like others, searched for evidence of keen competition between animals of the same species. He discovered, however, that the conception of nature as a field of unrestricted warfare was in reality a mere caricature of real life (1914, vii, ix):

> Two aspects of animal life impressed me most during the journeys which I made in my youth in Eastern Siberia and Northern Manchuria. One of these was the extreme severity of the struggle for existence which most species of animals have to carry on against an inclement Nature . . . And the other was, that even in these few spots where animal life teemed in abundance, I failed to find – although I was eager by looking for it – that bitter struggle for the means of existence among animals of the same species. (1914, vii)

Kropotkin's historical research indicated that struggles for existence were carried on not by individuals, but by groups of individuals cooperating with one another. This led him to reaffirm the predominance of mutual aid over competition in the quest for survival, and in the advance of civilization itself (1914, 129; 1924, 14).

In a later unfinished work entitled *Ethics* (1924), Kropotkin suggested that, apart from its role in evolution, mutual aid can provide "the germ out of which all the subsequent conceptions of justice, and . . . higher conceptions of morality" evolve (1924, 30, 61). He asserted that identification with the interests of a larger group, and not legal codes, best insures a basis for morality – that is, encourages people to treat each other equitably and to sacrifice some areas of personal freedom in the interest of the well-being of the group. Kropotkin used examples of relationships in the natural world to establish the basis for such a moral standard, obviating a need to appeal to faith alone or abstract metaphysical concepts. His optimism was such that he hoped these examples would help to reactivate a cooperative impulse in people, encouraging them to replace freedom-inhibiting institutions with those which support their cooperative inclinations.

The struggle for human expression

While Kropotkin recognized the importance of mutual aid in evolution, he also perceived that certain modes of political, economic, social, and spatial organization discourage its emergence. His studies of centralized institutions revealed the extent to which they remove from people the right to control crucial elements of their lives. Kropotkin found that centralized institutions inhibit the development of a cooperative personality, promote inequality, and limit economic progress and the satisfaction of pressing social needs.

The emphasis which Kropotkin placed on encouraging people to utilize all of their creative talents in cooperative endeavours led him to reject hierarchical capitalist divisions of labour. He identified the division of society into unequal classes, and the existence of extreme wealth beside extreme poverty, as the worst inhibitors of creative and communal expression (1898, 418–9). Capitalist work environments, which generate inequality, fragment work tasks, and create pyramids of power, transform work as a cooperative process into a competitive struggle. Extreme divisions of labour in time and space, and the separation of manual from mental skills, also stifle humane creativity (1892, 166). When labour is no longer an expression of social need, and its rhythms cannot be controlled directly by workers, a form of 'starvation of the spirit' arises which is as serious as material deprivation (1914, 242; 1892, 197–9; 1898, 364–73).

But Kropotkin did not believe that authoritarian work structures were the inevitable consequences of industrialization:

> . . . to speak of the natural death of the village communities in virtue of economical law is as grim a joke as to speak of the natural death of soldiers slaughtered on a battlefield. (1914, 28)

Associations of workers for cooperative purposes did not disappear naturally, but rather were systematically outlawed in places such as Britain as early as the fifteenth century (1914, 236). This led Kropotkin to conclude that discipline and extreme specialization are not required for the efficient functioning of either industry or agriculture, but are merely means by which the employer can better control the worker.

Anticipating contemporary theorists like Schumacher (1974), Kropotkin also identified the limitations of economies of scale, and the economic inefficiencies which can result from bigness and the over-concentration of economic activity in space (1898). Large factories inhibit production in small factories, not by maintaining a superior technical organization, but by monopolizing and dictating terms in the selling and buying markets. Concentration is 'the amalgamation of capitalists for the purpose of dominating the market and not cheapening the technical process' (1898, 354). Thus, whenever small industries can overcome these processes by association or some other means, they seem to succeed quite well (1898, 349).

Kropotkin also criticized governments, which historically utilize authoritarian relationships to inhibit cooperation and further economic and social privilege. Like advertisers, governments use their power to define reality for people – to sell their particular version of what is necessary, right, and valuable in life (1887, 241; 1914, 236–7). Bureaucracies also encourage people to give up personal involvement and a practical knowledge of social and economic affairs. Whether at the voting box, where 'representatives' are chosen infrequently, or in the classroom where children are encouraged to defer to 'experts', individual expression is limited, imaginative faculties pre-empted, and passivity encouraged. Kropotkin hoped to replace government manipulation of power with a 'politics' which aims at the direct participation of people in their daily lives.

He did not believe, however, that a free participatory society could function with adults who were subject to manipulation or who depended on experts for every important decision (1971, 53–61). Educational systems which spread imperialist ideologies, generate disrespect or lack of understanding for the complexities of other cultures, assign prescribed roles in an economic hierarchy, encourage the acceptance of authority, and stifle independence of thought, have to be eliminated (1971, 30–1; 1885, 950). Perceiving that the ultimate power of eduction is its ability to affect feelings of self-worth, Kropotkin saw a need to develop a libertarian approach to learning which would facilitate choice rather than mould character.

To the extent that spatial organization reinforces hierarchical social and economic relationships, it becomes another target for anarchist critique. Kropotkin identified the manner in which the criteria of social needs and ecological balance are ignored in determining land use under capitalism. He was also sensitive to the ways in which centralization and compartmentalization of space inhibit communalism and induce feelings of powerlessness in people (1914, 277).

In the process of developing a theory of 'mutual aid', Kropotkin thus laid the conceptual foundation for a radical theory of human ecology. He viewed nature and people in nature as organic, interrelated wholes — the actions of any one part affecting all other parts. Imbalances which exist in nature thus reflect imbalances which exist in human relationships. Kropotkin also considered people subject to many of the same processes as are found in the natural world, including evolutionary and, on occasion, revolutionary metamorphosis (Kropotkin, 1892, 81; Galois, 1976, 74).

To secure a balanced position within nature, Kropotkin urged people to uncover the laws of the environment and to act in accordance with them (1971, 141). He did not suggest that people refrain from intervening in nature, only that they base their intervention on a respect for, and understanding of, the natural world. Instead of trying to 'legislate' environmental awareness 'into' a citizenry, Kropotkin believed it was necessary to re-establish a sense of community and love of place. Rootedness in a particular environment would foster greater human interaction and a more intimate relationship with one's surroundings (1914, 1924).

Kropotkin did not believe that modification of the built environment could be a substitute for radical change in the fabric of social life. To eliminate the constraints which certain environments place on personal growth and communal expression, and to insure a supply of natural resources for the future, societies had to rid themselves of capitalism, the state, and the basic attitudes which underlie all hierarchical institutions. To harmonize the relationships between people and nature, it was necessary first to create a human community which lived in harmony with itself (Reclus, 1896).

Decentralism: the anarchist alternative

Although Kropotkin was continually impressed with the power of mutual aid to reassert itself even in the shadow of highly centralized institutions, he recognized that the maximization of freedom required vast alterations in the social and economic environment. A revolution of the magnitude necessary to produce such change necessitated a clear conception of alternatives, and a revitalization of human energy and creativity. Therefore, Kropotkin determined that revolutionaries had an obligation to provide much more than criticism of the status quo:

> Socialist papers have often a tendency to become mere annals of complaints about existing conditions. (This) . . . exercises a most depressing influence on the reader . . . on the contrary, . . . a revolutionary paper must be, above all, a record of those symptoms which everywhere announce the coming of a new era, the germination of new forms of social life, the growing revolt against antiquated institutions . . . to make one feel sympathic with the throbbing of the human heart all over the world, with its revolt against . . . injustice – this should be the chief duty of a revolutionary . . . It is hope, not despair, which makes successful revolutions. (1899, 418)

The aim of a revolutionary is not to create images of an artificial utopia or draw a 'master plan' for the future society. Efforts must be directed instead at identifying those tendencies for socialist expression which exist already in society, and then creating the space for those tendencies to be realized. Anarchist decentralism, as Kropotkin saw it, was thus a popular struggle to articulate a totally new way of life, to

> . . . reshape all relationships from those which exist . . . between every individual and his churchwarden or . . . station master to those which exist between neighbourhoods, hamlets, cities, and regions. (1970, 262)

Several of the books and articles which Kropotkin wrote between 1875 and 1914 set out some of the general principles upon which he believed an anarchist society and its institutions would be based. A brief sketch of this society appears in his autobiography as follows:

> . . . a new form of society is germinating in the . . . old one . . . a society of equals, who will not be compelled to sell their hands and brains to those who choose to employ them in a haphazard way, but who will be able to apply their knowledge and capacities to production . . . for procuring the greatest sum of possible well-being for all, while full, free scope will be left for every individual initiative. This society will be composed of a multitude of associations, federated for all the purposes which require federation: trade federations for production of all sorts, – agricultural, industrial, intellectual, artistic; communes for consumption, making provision for dwellings, gas works, supplies of food, sanitary arrangements, etc.; federations of communes with trade organizations; and finally, under groups covering all the country, or several countries, composed of men who collaborate for the satisfaction of such economic, intellectual, artistic, and moral needs as are not limited to a given territory. All these will combine directly, by means of free agreements between them . . . There will be full freedom for the development of new forms of production, intervention, and organization; individual initiative will be encouraged, and the tendency toward uniformity and centralization . . . discouraged . . . this society will be a living, evolving organism. (1899, 398–9)

At the core of this vision is a new definition of 'political economy' as a science devoted to the study of the needs of people and of the best means (or most favourable conditions) for satisfying those needs with the least possible waste of human energy (1892, 190, 194, 196–7). New criteria for production would incorporate social as well as economic priorities –

useful and interesting work as well as material security. These criteria would be set by each production unit in conjunction with the larger community, including its non-working members.

Kropotkin also envisioned a system of complete workers' and citizens' control. Subordinate/superior relationships would be replaced by direct producer/producer relationships or citizen association. Workers who exercise collective control over the means of production would learn many diverse aspects of the production and planning process, rotating jobs within the factory, and between factories, fields and communities. Wage systems which reinforce individualism and substitute material incentives for self-motivation would also be eliminated.

Kropotkin believed that all labour was 'social labour' and that it was impossible to individually measure the contribution of any one person (1892, 181). Every invention and instrument of production was the result of previous inventions. Every person as an equal inheritor from the past thus deserved equal claim to the goods of society (1892, 184; 1887, 249). This suggested that no limits be set on the consumption of abundant commodities and that a rationing system based on estimations of need be established to distribute scarcer commodities (1892, 186, 196).

Social revolution: from theory to practice

Kropotkin defined revolutions as 'periods of accelerated rapid evolution' which were as natural a part of human life as the slower periods of physical and social change (1899, 290-1). The implementation of revolutionary alternatives did not occur, however, as a result of the substitution of one political elite for another. Kropotkin, and other social anarchists, believed that creation of alternative institutions and modes of behaviour were the very essence of the revolutionary process, and that the organization and methods used to bring about revolution had to embody the same kinds of egalitarian, participatory relationships that the revolution was designed to create.

Revolutions of the type which Kropotkin described thus begin when people start to transform themselves into the kinds of individuals who can function without submitting to authority, and who can exercise responsible discretion over their lives. There is no anarchist party or intellectual elite and yet social anarchism requires organization – organization of workers and citizens into local and regional congresses to discuss current needs and problems (1899, 288-9, 402). Kropotkin believed that ordinary people were more than capable of understanding the root causes of their oppression and of considering modes of action (1899, 272-3, 395). Revolutionaries might help to awaken the initiative of local organizations by relating local problems to the larger social and economic system, or by investigating with them the practical aspects of a more equitable political form (1885, 954; 1899, 318, 323, 379, 402-3).

My experience is that when you speak to the Russian peasant plainly, and start from concrete facts, there is no generalization from the whole world of natural and social science that cannot be conveyed . . . *if you yourself understand it concretely.* (1899, xxiii)

The job of revolutionaries, then, was not to formulate doctrine or complex intellectual arguments, but to act as the 'midwives' to self-liberation. Their role was to communicate ideas, promote a spirit of inquiry, and involve as large a number of people as possible in thought processes and creative acts of change. They would facilitate the establishment of clear goals and methods, which would hopefully reduce the violence and mutual embitterment of a revolution (1899, 290–1; Keltie, 1882).

Kropotkin spent years condensing these ideas into pamphlets which workers could both understand and afford (1899, 275, 324, 423). A vast network set up at local, regional, national, and even international levels distributed these works as well as the newspapers of local work associations throughout Europe in the late nineteenth and early twentieth centuries. He hoped to inspire people to a new shaping of their lives by setting high ethical standards for his own behaviour and by utilizing every opportunity to demonstrate the value of social anarchism in practice (1914, foreword). Believing that the vision of a revolutionary society could be foreshadowed in miniature in a redesign of the personal behaviour and institutions of its aspirants, Kropotkin joined revolutionary circles composed of people described as 'closely united by their common object, and . . . broadly and delicately humane in their mutual relations' (1899, 317).

KROPOTKIN AS GEOGRAPHER

In the 1890s, the Royal Geographical Society of Britain held a banquet at which all the guests rose to honour Kropotkin for his contributions to the field of physical geography. Festivities began with a toast to the King. Consistent with his anarchist politics, Kropotkin refused to stand in tribute (Woodcock and Avakumović, 1950, 227).

This incident illustrates the strangely ambivalent relationship of 'respectability and disrepute' which Kropotkin maintained with the labourers of his profession (Baldwin, 1927, 22; Woodcock and Avakumović, 1950, 228). In spite of a close personal friendship with Scott Keltie (Secretary of the RGS from 1892 to 1915), Kropotkin declined the honour of being elected an official 'fellow' of the group. Hostile towards any organization under royal patronage, but committed to the advancement of science, he nevertheless established a close working relationship with its members (Kropotkin, 1971, 22; Woodcock and Avakumović, 1950, 227; Kropotkin, 30 January, 1892).

The personal support which Kropotkin received from an intellectual community which included such noted scholars as William Morris Davis, H. W. Bates, Patrick Geddes, and Victor Hugo, indicated the degree to which his unique contributions to science were appreciated. When imprisoned in France in 1882 for an alleged political bombing incident, hundreds of scholars, including professors from Cambridge, London, Edinburgh, St Andrews and Oxford Universities, leading officials of the British Museum, and editors of the Encyclopaedia Britannica, sought his release (Woodcock and Avakumović, 1950, 77–8, 81, 89, 194). Although it was rumoured that Kropotkin was offered a Cambridge Professorship of Geography on the condition that he abandon his anarchist revolutionary practices, most geographers seemed to ignore his political views (Woodcock and Avakumović, 1950, 228).

It is not difficult to understand why a basically conservative community of international scholars should have accorded Kropotkin such a high measure of esteem. In spite of increasing involvement in revolutionary politics, he maintained a strong commitment to his geographic research and scientific methodology. While most young revolutionaries in anti-Czarist Russia remained underground in the 1870s, Kropotkin openly attended a meeting of the Russian Geographical Society in order to explain his geological thesis on the Ice Cap. This led to his arrest and lengthy imprisonment in the terrible St Peter and St Paul fortress.

While imprisoned, Kropotkin wrote an immense two-volume work on the orography of Asia and on the glacial period in Finland and Central Europe (Keltie, 1921; Kropotkin, 1899, 349). He recalled in later years that the permission he received to read and write for brief periods during the day was the single factor which made incarceration bearable (Kropotkin, 1899, 350).

When imprisoned again in Clairvaux, France, in 1883, Kropotkin called upon his wife and associates to supply him with scientific reports (12 February, 1883, 2 March 1883, 3 April 1883). He also organized classes for fellow inmates in cosmography, geometry, physics, language and bookbinding (1899, 461).

In spite of incarceration and political activity, Kropotkin wrote several articles between 1893 and 1905 for *Le Révolté*, a radical European periodical, *The Times*, and such scholarly journals as *The Nineteenth Century*, *Geographical Journal*, and *Nature*. He also spoke frequently at geographical congresses and participated in lecture tours within Britain and the United States (1899, 424). He dealt with topics ranging from social commentaries on the condition of peasants in Russia and discussions of social anarchism, to technical discourses in natural science and analyses of Russian novelists. He was received enthusiastically by a wide range of audiences, from workers in union halls to fellow academics in prestigious institutions, such as the Lowell Institute in Boston ('Prince Kropotkin on Russia's Condition', 1901; 'Turgueneff and Tolstoi', 1901; 'Prince Kropotkin Talked on Anarchy', 1901; Kropotkin, 1899, 380–382; Woodcock and Avakumović, 1950, 227).

The purpose of geography

In the course of his studies, Kropotkin developed strong views on education in general, and geography in particular, many of which are set forth in a paper entitled *What geography ought to be* (1885). This paper, published as part of a larger report on geographical education in Britain, led to the creation of Royal Geographical Society lectureships at Oxford and Cambridge. It was, and still remains, a moving plea for the injection of social relevance into the content and methodology of geographic education.

Kropotkin believed that the general purpose of education was to create within people an awareness of the social forces acting upon them and a desire to resist political or social manipulation (1885, 954). He felt that geography was uniquely suited to serving these purposes because of its ability to capture the imaginations of children and promote a sense of mutual respect between nations and people:

> . . . the task of geography in early childhood (is) to interest the child in the great phenomena of nature, to awaken the desire of knowing and explaining them. Geography must render, moreover, another far more important service. It must teach us . . . that we are all brethren, whatever our nationality . . . geography must be – in so far as the school may do anything to counterbalance hostile influences – a means of dissipating . . . prejudice and of creating other feelings more worthy of humanity. It must show that each nationality brings its own precious building stone for the general development of the commonwealth, and that only small parts of each nation are interested in maintaining national hatreds and jealousies. (1885, 942)

Kropotkin believed that geography could emphasize the common bonds shared between workers across national boundaries (1885, 942). Because of its interdisciplinary nature, it could also provide a generalized conception of nature as a whole, and act as a great 'integrator of knowledge', (1899, 88–9):

> . . . combin(ing) in *one* vivid picture all separate elements of . . . knowledge; to represent it as a harmonious whole all parts of which are . . . held together by their mutual relations. (1885, 949)

Unfortunately, geography, as taught in nineteenth-century European schools, in no way resembled this ideal:

. . . the most attractive and suggestive (of sciences) for people of all ages – we have managed to make in our schools one of the most arid and unmeaning subjects. Nothing interests children like travels; and nothing is drier and less attractive in most schools than what is christened there with the name of geography . . . the harmonies of nature, the beauty of its forms, the admirable adaptations of organisms . . . all these may come later, but not in early childhood. The child searches everywhere for man, for his struggles against obstacles, for his activity . . . it is passing through a period when imagination is prevailing. It wants human dreams . . . tales of hunting and fishing, of sea travels . . . of customs and manners of traditions and migrations. (1885, 940–1)

Far from emphasizing the cultural and moral bonds between nations, European geography supported militaristic and imperialistic ventures.

Kropotkin believed that an alternative educational method could provide knowledge free from ideological domination. Education must be 'merged with apprenticeship', ideas derived from actual practice, and programmes of study based on direct community and workplace exploration (1898, 406; 1899, 117, 125). There must in the educational process be the combination of mental with manual labour, conception with execution (1898, 363–409). These beliefs led him to encourage educators to require students to work at reaching the truth rather than providing them with total explanations (23 March, 1893).

Decentralism and the patterning of an anarchist landscape

Kropotkin's geography provided a basis for, and was profoundly influenced by, his anarchist political beliefs. His focus was on the study of human interdependence for the purpose of sustaining life and furthering the evolution of new kinds of people. He proceeded through an examination of the relations between community, the primary cell of social life, and environment, its spatial and physical context. The idea was that certain kinds of human–environment relations and certain social and spatial modes of person-to-person interaction were conductive to the realization of the human potential. Kropotkin's geography searched for these relations in evolutionary history and the contemporary social reality.

Kropotkin's early work in physical geography laid the foundation for a broad love for and understanding of the natural world, and of people in diverse environments (Woodcock and Avakumović, 1950, 91):

The infinite immensity of the universe, the greatness of nature, its poetry, its ever throbbing life, impressed me more and more, and that never ceasing life and its harmonies gave me the ecstasy of admiration which the young soul thirsts for, while my favourite poets supplied me with an expression in words of that awakening love of mankind and faith in its progress which make up the best part of youth and impress man for a life. (Kropotkin, 1899, 97)

He disproved Humboldt's theory of the orographic structure of Asia, demonstrating that the great Siberian range ran not north to south or east to west, but northeast to southwest. He also initiated Arctic explorations in the 1870s, and developed theories on the glaciology of Finland. These activities enabled him to improve his powers of analytic thought, and increase his knowledge of scientific methodology. This concern for empirical verification, developed through his study of the physical world, carried over into his later consideration of social issues (Woodcock and Avakumović, 1950, 91). Kropotkin was the first to try and formulate a scientific basis for anarchism as a principle of life – to prove, using the methods of natural science, the negative effects of authority in all of its forms and the need for a complete reorganization of society on the basis of the free cooperation of independent associations (Kropotkin, 21 May, 1902: Kropotkin, 1971, 21; Kropotkin, 1899, 403). *Mutual Aid* (1914) and *Ethics* (1924) were attempts to establish a 'biological basis for freedom' – to prove that nature demanded that people be free in order for them to grow (Holloway, 1942, 35).

Kropotkin's social and economic geography illustrates the strong commitment he had to expanding the boundaries of conventional geographic study, and to combining intellectual interests with revolutionary intent. His environmental theory of 'mutual aid' justified the search for radical social change along anarchist lines. The socio-economic/spatial theory of 'decentralism' was then developed to describe the general principles upon which an anarchist society would be based.

The primary geographic and social unit in Kropotkin's decentralist society was the self-governing commune. This was a 'natural community', formed by groups of urban and rural dwellers, for the purpose of apportioning common property to a variety of social and economic uses (Guerin, 1970, 57). Kropotkin anticipated contemporary proponents of decentralization by suggesting that there are natural limits to the size of a self-governing group that wishes to encourage the full participation of its members. He was reluctant, however, to suggest an ideal figure for community size.

The aim of decentralization in an anarchist society is to increase the potential for human interaction, and to avoid cultural isolation (Kropotkin, 1914, 277). Kropotkin thus suggested that within cities, single core areas be replaced by a multitude of cores, each reflecting a unique subculture. Land use within these areas should be desegregated so as to integrate living, working, and recreational space. Distances between workplace and residence should also be decreased and land cultivated within city limits (1892, 103–4). These spatial modifications would decrease the dependence of cities on rural areas for food, and climinate the unnatural separation of work from the rest of social life. Kropotkin believed that the integration of a variety of activities in space would complement changes in the social relations of production, creating more diverse environments and allowing for a more flexible use of time (1914, 151–222).

Kropotkin also foresaw a need for the expropriation of private property and capital (1892, 177). He regarded urban infrastructure, including roads, houses, and transportation arteries as community property, because generations of workers were responsible for their creation. He viewed towns as:

> . . . agglomeration(s) of thousands of . . . houses, possessing paved
> streets, bridges . . . and fine public buildings . . . offering . . . a
> thousand comforts, . . . a town in regular communication with
> other towns, and itself a centre of industry, commerce, science and
> art, a town which the work of twenty of thirty generations has made
> habitable, healthy and beautiful. (1914, 105)

> What would be the value of an immense London shop or storehouse
> were it not situated precisely in London which has become the
> gathering spot for five millions of human beings? (1887, 249)

Revolutionary schemes for land redistribution must therefore be based on the criteria of need rather than ownership.

At the regional level, Kropotkin favoured self-sufficiency in the provision of essential commodities in order to diminish the dependence of one region on another, eliminate gross inequalities, and enhance the prospects for inter-regional cooperation. The aim was not to eliminate *all* forms of specialization, only those 'tyrannies' which accompanied *over*-specialization. The gradual disappearance of a sharp dichotomy between agricultural and industrial land use would also prevent the overconcentration of population, and enable local areas to become more attuned to the needs of inhabitants (1914, 100). These ideas anticipated much later work by Ebenezer Howard (1965). Lewis Mumford (1961), Patrick Geddes (1968), and Paul and Percival Goodman (1947), except that Kropotkin did not believe that spatial alterations could become a substitute for social revolution.

Kropotkin rejected the pastoral visions of Tolstoy and the opposite extreme of centralized planning, on the ground that each alternative was too simplistic for a modern complex society. Instead, he advocated a form of regionalism which would promote the active interchange of information and products between areas via multiple federations (1893; 1898). Economic federations, an idea derived from Proudhon's earlier work (1851, 104–5), would consist of associations of workers by separate industry and home-area. These associations would contract with each other on a temporary basis for the provision of goods and services. The integration of small-scale economic activity between and within regions would then produce economies of scale superior to those formerly produced by large industry. Geographic federations formed by self-governing communes at the local, regional, and national levels would facilitate exchange, and coordinate plans requiring the cooperation of several areas. 'Social communes', or extra-territorial linkages of people interested in sharing cultural and scientific ideas, would also be created (1892, 61–71). In each federation the major decisions would be made at the local level. Regional and national councils would be responsible only for the coordination of locally-determined activities.

The space-economy of a hypothetical Kropotkinian landscape would thus be characterized by little distinction between areas of resource extraction, manufacture, and consumption. Information, products, and labour would circulate freely, causing regions to experience a simultaneous increase in the concentration of some economic functions and the dispersal of others. Exchange of speciality items would continue between regions, with a probable decline in the exchange of necessities. Price and cost indicators, the main criteria for location under capitalism, would be replaced by a local community's evaluation of its needs and the resources available to it, either locally or by exchange. Finally, the whole layout of transport systems would alter as networks connecting communities to each other replace hierarchies connecting them to some central point in space.

Many of Kropotkin's ideas for the social and spatial reorganization of society on a decentralist basis were applied in Spain in the late nineteenth and early twentieth centuries, when anarchism emerged as a powerful movement for radical social change. Inspired by Kropotkin's writings, a massive social revolution accompanied the Civil War in 1936.

When thousands of workers and peasants collectivized the land, factories, transportation networks and public services in July 1936, a capitalist mode of production was replaced by direct workers' and citizen control. New irrigation systems allowed peasants to bring additional land under cultivation, expanding and diversifying production. Social landscapes were altered to accomodate new educational, cultural, and health facilities. Exchange between collectives on a local and regional scale also extended transportation and health services into areas that had never been serviced before. In short, a revolution which began by altering social relationships created totally new spatial formations (Breitbart, 1978).

CONCLUSION

Kropotkin left England in 1917, at the age of seventy-five, for Russia, where he hoped to see his anarchist ideals realized (Prince Kropotkin's Farewell, 1917). He died, disappointed, in a small town near Moscow four years later. Numerous obituaries reflected the wide appeal which he had as a scientist, humanitarian, and revolutionary. In one particularly moving tribute, Paul Goodman identified the chief lesson which Kropotkin provided for young people: how an authentic professional becomes a revolutionary and lives his/her life as 'pure social action'. Kropotkin understood that most intellectual and scientific pursuits were, by definition, ideological and wrought with social effects. By just trying to pursue his profession 'with courage and integrity' he thus found it necessary to seek revolutionary change in society (Goodman, 1968 (1899), xxi–xxix).

This view lies in sharp contrast to the view taken by professional geographer Scott Keltie in an obituary written for *The Geographical Journal* in 1921. In this article, Kropotkin's political actions and social conscience are seen as unrelated, and a hindrance to, his scientific research:

> The announcement of the death of Prince Peter Alexeivich Kropotkin on 8 February, . . . will have been received with regret by a wide circle of all classes and all creeds . . . This is not the place to deal in detail with Kropotkin's political views, except to express regret that his absorption in these seriously diminished the services which otherwise he might have rendered to geography. (Keltie, 1921, 316–9)

One point over which there is little dispute concerns Kropotkin's reluctance to observe scientific fact with cold objectivity (Robinson, 1908, 88, 111; Keltie, 1921, 319; *Centennial Expressions*, 1942). As Malatesta remarked, 'In him the heart spoke first and then reason followed to justify and reinforce the impulses of the heart'(Richards, 1977, 264).

That Kropotkin tried to revolutionize the discipline of geography in a number of key areas is indisputable. His theories of geographic education, human ecology, and decentralization were clearly aimed at halting the use of geographic research for exploitative and imperialistic purposes. Instead of subscribing to a narrow physical view of the discipline, Kropotkin emphasized the interrelatedness of natural and social processes, and the importance to both of cooperation over competition. He also recognized that a true ecology movement had to be linked to a revival of community, and hence, radical political, social, and economic change.

The contemporary relevance of Kropotkin's ideas is vast. With regard to current struggles for social change, he pointed out the importance of connecting means to ends, process to form. In the political spectrum, this led him to favour grass-roots organizing whereby powerless people reject an externally-imposed design for living, and begin to transform themselves into active agents for change. In professional arenas, he decried the dehumanizing effects of bureaucracy and social engineering. This led him to favour elimination of the lines which separated those 'who plan' from those who are 'planned for' (Goodman, 1968 (1899), xxii).

Were Kropotkin alive today, he would surely want to distinguish between rural escapism and true decentralization, garden cities and 'natural communities', environmentalism and social ecology, illusions of liberation and true forms of self-management. He would challenge our 'sacred' notions of economic efficiency, demonstrating that current technologies and work organization depend not upon neutral scientific principles, but upon mode of production and property interests. Kropotkin would also suggest that it is possible to enhance the potential for innovation, equity, and efficiency through cooperative behaviour – that diversity and social interaction are as important as autonomy in promoting personal development – and that smallness of scale combined with self-management and federalism provide a basis for the integration of local and regional, economic and social interests.

Kropotkin devoted his life to the elimination of those institutions which overtly dominate peoples' lives or inadvertently cause the atrophy of the desire of people to mould their environments. He believed that a revolution which began by initiating change in social and economic relationships would thus end up creating a totally new environment. Decentralism was the revolutionary philosophy and mode of socio-spatial organization which he felt would comprise the backbone of a new cooperative mode of existence. It was both the geography and sociology of anarchism.

Kropotkin never drew a precise blueprint for a new society, believing that each group, and even each individual within a group, must experiment to find its particular version of the anarchist ideal. The broad objectives of his decentralist vision were nevertheless clear: to achieve a synthesis between personal freedom and social responsibility. Freedom to develop the self while acknowledging the freedom of others, freedom to act independently while not damaging others, freedom to grow and continually renew oneself in a changing environment, these were the main components of Kropotkin's anarchism and his geography.

With these achievements to his credit, one must ask why Kropotkin has been almost thoroughly neglected by his discipline. The early neglect of his work in social and economic geography may perhaps be explained by the fact that these ideas were developed during a period of increased centralization when capitalist nations were attempting to consolidate further their power over world resources. The only reasonable explanation for the persistence of this neglect is that many of the same conditions prevail today. Thus, the 'geographic revolution' which Kropotkin initiated one hundred years ago was, and still remains, tied to an unrealized social revolution. Kropotkin's geography, like his anarchism, continues to present an enormous challenge to the professional and political status quo.

References

Breitbart, M. 1975: Impressions of an anarchist landscape. *Antipode* 7, 44 -50.
Breitbart, M. 1978: *The theory and practice of anarchist decentralism in Spain, 1936–1939: The integration of community and environment.* Ph.D. dissertation, Clark University.
Centennial Expressions on Peter Kropotkin, 1842–1942. 1942: (New York).
Darwin, C. 1859: *On the origin of species* (1964: Cambridge, Mass.).
Galois, B. 1976: Ideology and the idea of nature: The case of Peter Kropotkin. *Antipode* 8, 1–16.
Geddes, P. 1968: *Cities in evolution* (New York).
Goodman, P. 1968: Preface. In Kropotkin, P. 1899: *Memoirs of a Revolutionist* xxi–xxix, (New York).
Goodman, P. and P. 1947: *Communitas* (New York).
Guerin, D. 1970: *Anarchism* (New York).
Hoffman, R. 1972: *Revolutionary justice: the social and political theory of P. J. Proudhon* (Urbana, Illinois).
Holloway, W. 1942: What Kropotkin means to me. In *Centennial Expressions*, 34–5, (New York).
Howard, E. 1965: *Garden cities for tomorrow* (Cambridge, Mass.).
Keltie, J. S. Letter to editor of *Pall Mall Gazette*, 22 November 1882. (Royal Geographical Society, London).
Keltie, J. S. 1921: Obituary for Peter Alexeivich Kropotkin. *The Geographical Journal* 57, 316–9.
Kropotkin, P. Letter to J. Scott Keltie, 12 February 1883 (Royal Geographical Society, London).
Kropotkin, P. Letter to J. Scott Keltie, 2 March 1883. (Royal Geographical Society, London).
Kropotkin, P. Letter to J. Scott Keltie, 3 April 1883. (Royal Geographical Society, London).
Kropotkin, P. 1885: What geography ought to be. *The Nineteenth Century* 18, 940–56.
Kropotkin, P. 1887: Scientific bases of anarchy. *The Nineteenth Century* 21, 238–58
Kropotkin, P. 1892: *The conquest of bread* (1972: New York).
Kropotkin, P. Letter to J. Scott Keltie, 30 January 1892 (Royal Geographical Society, London).
Kropotkin, P. Letter to H. R. Mill, 23 March 1893 (Royal Geographical Society, London).
Kropotkin, P. 1895: The present condition of Russia. *The Nineteenth Century* 38, 519–35.
Kropotkin, P. 1898: *Fields, factories and workshops* (London).
Kropotkin, P. 1899: *Memoirs of a revolutionary* (1968: New York).
Kropotkin, P. Letter to J. Scott Keltie, 21 May 1902 (Royal Geographical Society, London).

Kropotkin, P. 1902: *Mutual aid: a factor of evolution* (Boston, Mass.).
Kropotkin, P. 1919: Direct action of environment and evolution. *The Nineteenth Century* 85, 70–89.
Kropotkin, P. 1924: *Ethics: origin and development* (New York).
Kropotkin, P. 1970: *Selected writings on anarchism and revolution*, Miller, M. editor (Cambridge, Mass.).
Kropotkin, P. 1971: *Revlutionary pamphlets*, Baldwin, R. editor (New York).
Miller, M. 1976: *Kropotkin* (Illinois).
Mumford, L. 1961: *The city in history* (New York).
Prince Kropotkin on Russia's condition. 1901: *New York Times* 30 March.
Prince Kropotkin talked on anarchy. 1901: *New York Times* 1 April
Prince Kropotkin's farewell. 1917: *The Times* 8 June.
Proudhon, P. 1851: The general idea of revolution in the 19th century. In Shatz, M., editor *The essential works of anarchism* (New York: 1971).
Reclus, E. 1896: The progress of mankind. *Contemporary Review*, 70.
Richards, V., editor, 1977: *Malatesta: life and ideas* (London).
Robinson, V. 1908: *Comrade Kropotkin* (New York).
Schumacher, E. H. 1974: *Small is beautiful* (London).
Stirner, M. 1845: The ego and his own. In Shatz, M., editor, *The essential works of anarchism* (New York).
Stoddart, D. R. 1976: Kropotkin, Reclus, and 'relevant' geography. *Bulletin of Environmental Education* 58, 21.
Turgueneff and Tolstoi: Prince Kropotkin's lecture on the Russian novelists. 1901: *New York Times* 31 March.
Woodcock, G. and Avakumović, I. 1950: *The anarchist prince: a biographical study of Peter Kropotkin* (London and New York).

9

Elisée Reclus, an Anarchist in Geography

G. S. Dunbar

Elisée Reclus (1830–1905) was perhaps the most prolific geographer who ever lived. I have not counted his actual word output and compared it with that of possible rival claimants, such as Alexander von Humboldt or William Morris Davis, but I think that Reclus's production was greater. Humboldt had the advantages of greater wealth (and therefore leisure), fewer family duties, and a longer lifespan than Reclus. Davis was driven to high productivity by the Puritan work ethic and, initially, by job insecurity. Reclus was a disciplined, facile writer who could devote nearly all his waking hours to reading and writing. His prodigious literary output was due mostly to the simple fact that that was the way he made his living. Most professional geographers have earned their livelihood as university teachers, but Reclus was supported by his pen. Although he was a professor for the last eleven years of his life, he was not paid for teaching, and he had to rely on the sale of his articles and books. This was a rather ironic circumstance for a professed anarchist, and Reclus often resented the necessity of writing for a largely bourgeois audience instead of devoting his attention to the dissemination of anarchist propaganda.

BIOGRAPHICAL SKETCH

Elisée Reclus was born on 15 March 1830 in Sainte-Foy-la-Grande, a village on the Dordogne River in southwestern France, the third child of Jacques Reclus, an impecunious Protestant pastor, and his wife Zéline. Of the numerous Reclus children, eleven lived to full maturity. Although Elisée spent most of his youth away from home, he always felt close to his family, and he was remarkably similar in character to his father. The mature Elisée professed to be an atheist and anarchist, but his principles were virtually identical to those of his father, who remained a devout Christian. Actually, Jacques Reclus could be described as an anarchist in the same way that Christ and the early Church Fathers were anarchists.

The Reclus family moved to the Orthez region in 1831 and remained there for the next half-century, but Elisée spent his early childhood with his maternal grandparents in another village. He followed his older brother and sister to a Moravian school in Neuwied, Germany, in 1842, but returned to his native village to complete his baccalaureate in 1848. After a year in a Protestant seminary and another year teaching French in his old school in Neuwied, Elisée studied theology for six months at the University of Berlin in 1851. At Berlin he quite fortuitously attended the popular lectures of Carl Ritter in geography, and this set him on his life course. Reclus subsequently fell away from religion, but his geography was not greatly different from that of the devout Ritter.

In September 1851 Elisée and his brother Elie returned to Orthez, where they resisted the coup d'état of Napoleon III on 2 December. They were not officially banished or even reprimanded by the authorities, so far as I can determine, but they chose to leave France before any action could be taken against them. They arrived in London on New Year's Day 1852 and found tutoring jobs. After a brief stint as a farm manager in Ireland, Elisée sailed to Louisiana, where he was employed as a tutor to the children of a sugar planter in the period 1853–5. After 18 months in Colombia, where he vainly attempted to carve out a small farm, Elisée returned to France in 1857. He lived in Paris for the next 14 years, employed from December 1858 onward by the Hachette publishing house, where he worked on travel guides and other geographical publications while playing an active rôle in the Paris Geographical Society.

In the 1860's Elisée and Elie furthered their socialist educations and came under the influence of Michael Bakunin, the foremost anarchist of the day. They joined Bakunin's secret fraternity and followed him into the International Workingmen's Association. The Reclus brothers were not true anarchists at that time, however, because they still subscribed to democratic and republican principles. They were thoroughly radicalized only after they joined the ill-fated Paris Commune in March 1871. Elisée was captured in the first days of the fighting, but Elie escaped to Switzerland. Elisée spent the next 11 months in prisons in and around Brest and Paris. His sentence of deportation to New Caledonia was commuted to 10 years of banishment, and he chose Switzerland as his land of exile, partly to be near Elie and partly to remain in close contact with his publishers. Hachette had already published several of his books, including the two-volume *La Terre* (1868–9) that established his international reputation in geography, and they now (1872) agreed to publish a five- or six-volume world geography. This great work, the

F

Nouvelle géographie universelle, actually appeared in 19 volumes between 1876 and 1894. Although amnestied in 1879, Reclus continued to live in Switzerland until 1890, when he returned to Paris. His 18 years in Switzerland were almost completely occupied by the exacting task of preparing the great encyclopaedia and also by his anarchist activities. He can be said to have crossed over from his socialist republican ideas into true anarchism by 1876, although some writers think that they see indications of his anarchist tendencies as early as 1850. In any event, he made a public profession of his new faith in 1876, just before the death of Bakunin, and in February 1877 he allied himself with another geographer-anarchist, Peter Kropotkin (see chapter 8), in a union that was fruitful to both – fruitful not only in enhancing their geographical and anarchistic writings but in achieving a blend of the two.

On the completion of the *Universal Geography* in 1894, Elisée Reclus moved to Brussels and remained there for the last 11 years of his life. He had been promised a position in the Free University of Brussels, but university officials reneged after anarchist violence broke out in France in December 1893, and so Elisée, his brother Elie, and some like-minded professors and students founded the New University of Brussels, which coexisted peacefully with the Free University for the next 20 years. The work of the New University did not receive government accreditation, and professors were not paid for their teaching, but Elisée established a Geographical Institute in the University in 1898, and he managed to support a small staff by the revenues from publications and cartographic ventures. A major preoccupation of his Brussels years was his scheme of building a huge relief globe, either in connection with the Paris Exposition of 1900 or other world fairs, but this project never came to fruition (Dunbar, 1974). Elisée Reclus died in Thourout, Belgium, during the night of 4–5 July 1905 and was buried with his brother Elie in Ixelles. He was survived by his third wife, a daughter, and numerous other relatives. His sister Louise, who acted as his secretary in Brussels, was instrumental in the gathering and ordering of his papers, and his nephew Paul, Elie's son, directed the Geographical Institute until the New University was closed in 1914.

ANARCHISM

Many people, including perhaps the majority of geographers, think of geography and politics as two entirely separate compartments, and to them an anarchist geographer might seem to be a contradiction in terms or, at best, a curious mésalliance. I hope to show that anarchism and geography can be a logical combination; at least they were for Elisée Reclus and his friend Peter Kropotkin. A reader of their geographical works will be disappointed if he expects to find in them an explicit statement of their anarchism, for in practice these anarchist geographers were scarcely different from the mainstream geographers of their day. Actually, Reclus and Kropotkin could both be considered as mainstream geographers because of the perennial interest in their work and the timelessness of their message. It might be profitable to think of the field of geography as something like a braided stream, where the main channel is often difficult to find at any particular time but where several anastomosing channels are doing the work.

The term 'anarchy' simply refers to the state of being without rule or government. Although some people might dream of building a society that has no governmental structure whatsoever, as a practical matter most anarchists would settle for a highly decentralized society in which there are no repressive controls. In its simplest form anarchism is scarcely different from the Christian fundamentalism of Pastor Jacques Reclus, so it was not really a radical step for his son Elisée to give up his old faith for a new one. Anarchism and Marxism function in many respects like religions. They may not have an explicit belief in a supreme deity or in a heavenly paradise, but they certainly possess their saints and dogma. They are utopian and millennialist in much the same way as their sister sects of the nineteenth century. Reclus and his comrades were influenced by the French utopians, such as Saint-Simon, Fourier, and Cabet but, despite temptation, they did not subscribe to the utopian schemes of planting independent communities in the New World. As Reclus said, 'Never will we separate ourselves from the world to build a little church, hidden in some vast wilderness' (Reclus, 1884, 637; 1900, 1–2).

Elisée Reclus shared Kropotkin's principles of anarchist communism, according to which the product, as well as the means of production, would be held in common and distributed according to need. Individualism would be encouraged, but extreme individualism, which often manifested itself in violent forms of expression in the 1880s and 1890s, was an embarrassment to the idealistic anarchists such as Reclus, although he could not bring himself to disown the fanatics who committed violent acts in the name of anarchism. Jean Maitron, the leading French authority on the history of anarchism, has referred to Reclus's brand as 'sentimental anarchism', as compared with the 'educationist' anarchism of Jean Grave, the federalist and revolutionary anarchism of Peter Kropotkin, the eclectic anarchism of Sébastien Faure, the insurrectional anarchism of Charles Malato, the aesthetic anarchism of Jacques Mesnil, and the socialist and proletarian anarchism of Errico Malatesta. I do not pretend to understand all the nuances of meaning, and, even if I did, there is not space enough here to dissect them. These categories are not, of course, mutually exclusive, for most of them could be applied to Reclus's philosophy, which was so wonderfully tolerant that he could embrace nearly everyone on the far left side of the political spectrum. His warmth and tolerance, plus his greater age (by the 1890s there was a considerable gap between Reclus and most of the other anarchists), caused him to be venerated and to be granted considerable immunity from criticism. He became a father figure to the ardent youth, and he was even described, not derisively, as 'the pope of the sect' (Woodcock and Avakumović, 1950, 317; Maitron, 1951, 418).

POSSIBLE INFLUENCES OF RECLUS'S ANARCHISM ON HIS GEOGRAPHICAL WRITINGS

An unprepared reader might search in vain through the vast corpus of Reclus's geographical works for manifestations of his political sentiments. His writings abound in rhetorical passages that might be interpreted as subtle anarchistic passages, but those passages might just as well have been written by utopians, Comtian positivists, or even theologians. His nature sketches, for example – Histoire d'un ruisseau (1869) and Histoire d'une montagne (1880) – contain vaguely worded statements about liberty and the unity of mankind, but what person would not subscribe to such idealistic or humanitarian sentiments? In fact, these books might have been conscious imitations of Jules Michelet's popular nature sketches of the 1860s. Michelet was something of a rebel himself, though not an ultraist like Reclus, and the latter was careful to distinguish Michelet's views from his own more optimistic ones. Optimism is not, of course, a monopoly of anarchists, and in Reclus's case it would appear to derive from his early religious training. In any event, one cannot isolate passages of Reclus's geographical works and state categorically that they could only have been written by an anarchist. In truth, his works are not greatly different from those of the other geographers of the day – geographers from all parts of the political spectrum. Although Max Nettlau, Reclus's friend and biographer, has stated that only an anarchist could have conceived and carried out Reclus's monumental Nouvelle géographie universelle (Nettlau, 1928, 183), his point is perhaps unarguable but still highly doubtful because one can think of so many undisciplined anarchists and disciplined non-anarchists. Still, it makes one think of such writers as the two American hardy perennials, Lewis Mumford and Will Durant, who both received an early exposure to the works of Elisée Reclus and Peter Kropotkin and who have produced powerful and coherent bodies of work that now span six decades.

Although his apologists, and even Reclus himself late in life, have claimed that the officials of the Hachette publishing house suppressed his anarchism and gave him careful instructions not to offend his bourgeois readership, it is quite clear that Reclus was given virtually free rein in writing his geographical works. His editor, Emile Templier, never censored Reclus's work; indeed, after making largely stylistic suggestions about the first sections of the *Universal Geography*, he never even bothered to read the material before it was published. Reclus did not need a copy editor, and his punctuality could be relied upon absolutely. Templier was more concerned with the possible effects of Reclus's initially colourless writing style upon the readers than with their reactions to any of his political commentary.

Reviewers of the *Universal Geography* found numerous examples of what they thought were Reclus's prejudices, but they did not detect any underlying anarchist philosophy. A French reviewer complained that Reclus had under-estimated the number of French-speaking people in the world and over-estimated the German speakers – certainly a touchy topic in the 1870s – and he also felt that Reclus showed too much admiration for the Turks as opposed to their old Christian foes (Gaidoz, 1875, 362–3; 1876, 37). A British reviewer in *The Times* noted 'a little soreness . . . in speaking of Germany, and especially of Alsace-Lorraine.' In a review of the volume which included the British Isles, the *Times* writer commented on the 'distinctness and accuracy wonderful in a foreigner and a Frenchman.' 'M. Reclus touches with no light hand our social sores, but altogether, and considering his extreme political opinions (which, however, never obtrude themselves in this work), we have no reason to be dissatisfied with his sketch of ourselves.' In a review of Volume 8 (India and Southeast Asia) the same reviewer noted that 'M. Reclus's animadversions on the conduct of the English in India are not more severe than those which have been indulged in by English writers and publicists themselves . . . he admits that English statesmen have themselves foreseen all the difficulties he presents'(*The Times*, 1877; 1879; 1883). These were reviews of the original French edition of Reclus's work. The English translations that appeared later were always slightly irksome to Reclus because the translator, E. G. Ravenstein, whether unwittingly or not, sometimes managed to blunt the thrust of Reclus's statements. British bias can be detected in some passages, as when the translator dealt with Reclus's statement, 'Great Britain . . . has taken over Spanish Gibraltar . . . and Italian Malta', but left out the words 'Spanish' and 'Italian'. Some of the translations were simply erroneous, as in substituting 'Lithuanians' for 'Letts', and others were inept, as in the following passage which was rendered incomprehensible in translation: 'The geographers of the Renaissance liked to portray her [Europe] as a crowned Virgin [Mary]', which came out as 'The geographers of the period of the revival of letters compared it to a crowned virgin.' When Reclus reported that each Bosnian consumed on the average 130 litres of slivovitsa (plum brandy) a year, the translator converted the 130 litres to 34 pints and then gratuitously added the qualifier 'detestable' to describe the drink! Except for the conscious deletion of anti-British sentiments, I think that the translator's gaffes were honest errors rather than editorial changes (Reclus, 1876, 20, 30, 50, 205; Reclus, 1883, 13, 20, 34, 131).

Like so many Europeans, Reclus had a sort of love-hate relationship with the United States. Although he had spent more than two years there – albeit in Louisiana, an atypical American region – and later returned for two brief visits in 1889 and 1891 to collect data for the relevant volume of the *Universal Geography*, he viewed the country from a decidedly European and leftist stance. He shared the ambivalent views of many European intellectuals who saw in America the promised land but who castigated Americans severely for not suceeding where Europeans had failed. European socialists hoped that the inevitable world-wide revolution would begin in prosperous America rather than in some backward country like Russia (Moore, 1970). Evidence of social unrest or deteriorating economic conditions in the United States was avidly seized upon by the Europeans, who would interpret it as a portent of long-term trends and not as a singular occurrence or ephemeral phenomenon in what was essentially still a frontier country. Reclus took keen interest in pauperism and the rural problems of overproduction and outsized farms. For example, he was one of many European leftists who thought that they perceived the beginnings of a debilitating factory system in the wheat farming of the Red River Valley in North Dakota and Minnesota. There were large-scale farming enterprises elsewhere, to be sure, but the 'bonanza' farms near the Red River provided a striking example of capitalistic monopoly and oppression. European peasants were warned that the success of these large farms might cause European landlords to reorganize along similar lines. Who could foresee in 1880, when Reclus made his dire predictions, that the bonanza farms represented only a temporary stage in the evolution of the wheat frontier and that they would be replaced by smaller, owner-operated farms after the turn of the century? The fluidity of the land and labour situation in North America seemed to escape Reclus, who had referred to bonanza farming as 'a triumphant example of what can be done by large-scale scientific agriculture but a no less striking example of the monopoly which a few capitalists can assume over the work and the life of all' (Reclus, 1880, 1; compare Drache, 1964).

Reclus's most ample demonstration of his anarchist geography was in his six-volume work *L'Homme et la terre*, which was mostly published posthumously. He referred to this work as a 'social geography' – an interesting early use of a term that has gained great popularity in the 1960s and 1970s (Dunbar, 1977). *L'Homme* is a broad-canvas historical work that impresses the reader with its sweep and verve. As the volumes progress from ancient times to modern, their geographical content declines, and the discussion of contemporary institutions is quite journalistic in tone. The work is of interest today for the anarchist's view of history, with emphasis on those matters that are of perennial concern to the *engagés* – social ills and their solutions. In *L'Homme* Reclus reiterated his concerns about the unequal distribution of wealth and underscored the importance of the study of geography in making an inventory of the world's resources and suggesting a plan for their equitable spread.

Reclus would argue that the basic decisions about production and consumption should not be made by bureaucrats but by communities or associations of free labourers who have transcended avarice and the lust for power. The goal of 'free production and equitable distribution' is attainable, and Reclus was convinced, in his historical survey, that mankind was definitely making progress toward that happy state. His undying optimism suffuses the six volumes of *L'Homme* and gives them their unifying thread.

Reclus was opposed to the Neo-Malthusianism of many French anarchists, such as Paul Robin, because he believed that it was maldistribution of resources, rather than overpopulation, that was responsible for the world's ills. According to his calculations, hunger and poverty were caused by mismanagement, because the earth was producing twice as much food and three times as many industrial products as are required (Reclus, 1887). Kropotkin shared Reclus's views on population and production, but he feared that such an optimistic view of the earth's productivity might encourage complacency or laziness (P. Reclus, 1964, 116).

RECLUS'S INFLUENCE ON GEOGRAPHERS

When geography, history, and the other academic disciplines underwent a rapid revitalization and professionalization in the early years of the Third Republic, Elisée Reclus did not play a direct part because he was living outside France and because he lacked the doctorate that was necessary for obtaining a professorship in a French university. He perhaps could have been appointed to a chair in the Collège de France or a Swiss university, but when he was finally offered a post, it was in a Belgian university. Brussels could not provide a power base such as that which a professor in the University of Paris would have, but it gave him easy access to Great Britain and especially to his two most important contacts in the field of geography – Peter Kropotkin and Patrick Geddes.

From their first meeting in 1877 until Reclus's death 28 years later, Kropotkin and Reclus were firm friends and close professional associates. The influence of each on the other's work was great but incalculable. Kropotkin gave Reclus ample help on the sections of the *Universal Geography* that dealt with Siberia and Eastern Europe, and Reclus supplied French titles for Kropotkin's works. Geddes was close to Reclus in the last decade of the latter's life, and he aided considerably in the enlargement of Reclus's reputation in Britain. Geddes encouraged Reclus's cartographic schemes, notably the Great Globe, but most of these projects were never completed. Geddes was a libertarian, though not a true anarchist, and he was an effective interpreter of his friend's views. He spread his own and Reclus's style of geography among geographers, such as A. J. Herbertson, and non-geographers, such as Lewis Mumford.

Reclus's influence on contemporary French geographers was less direct, however. They had all read his writings and generally approved them, but they did not communicate with him or interact closely. This aloofness was not due to any political differences but simply to the fact that Reclus lived outside France, except for the period 1890–4, and he chose not to join in the construction of the newly emerging professional field of geography. The academic geographers were all younger than Reclus, and they paid homage to the master without subjecting his work to searching scrutiny. They credited him with laying the foundation for the New Geography, but the superstructure was of their own making. Reclus was an active member of the Paris Geographical Society before the Franco-Prussian War but not afterwards. The geographers of Switzerland seemed to get more of Reclus's attention, both during and after his residence there, and he contributed numerous articles to the *Bulletin* of the Neuchâtel Geographical-Society. The Belgian geographers enjoyed his presence among them, but, again, like the Swiss, they constituted a relatively small and unimportant body.

In a history of French geographical thought covering the period from 1872 to 1969, André Meynier asked why Elisée Reclus is not better known today, and he suggested that three factors might be responsible for Reclus's fading reputation: his 'Marxist' (*sic*) ideas, his long years of exile, and his easy-going but undemanding writing style. Reclus was not, of course, a Marxist, although he shared many of Marx's basic ideas. His advanced political opinions did not harm his reputation with his largely bourgeois readership, in his own or succeeding generations. Reclus's years of true exile, 1872–9, were highly productive, although Meynier was correct in implying that he could not have had the same influence as a Parisian professor while operating from a base in Switzerland or Belgium. Reclus's writing, said Meynier, had the tone of a pleasant conversation and lacked the didactic or structured style that the French, being car-tesians, seem to prefer. He credited Reclus with creating a favourable atmosphere for geography but not with a commanding rôle in directing the course of its evolution (Meynier, 1969, 13).

In recent studies that are highly sympathetic to Elisée Reclus, the French geographer Béatrice Giblin has said that Reclus has been 'erased' from the history of geography in France because people were afraid of his anarchism and because he was not in the mainstream of the new university-centred geography. The latter explanation would be more admissible than the former, but, even more basically, I do not accept the initial premise that Reclus has been erased. Obviously, his fame diminished after his death, but that was due simply to the passage of time and not to conspiracy. French geographers have not forgotten Reclus. They seem to have placed his life in proper perspective, neither unduly glorifying nor denigrating his achievements. There is no reason to suspect that Reclus's geography has been suppressed out of fear. Dr Giblin thought that Reclus's opposition to colonialism was not total, because he appeared to make an exception of Algeria, where one of his daughters was living with her family, but Reclus was actually rather critical of the French administration in Algeria (Giblin, 1971, 4–6, 115, 184, 200–206, 219–33; 1976, 30–49).

CONCLUSION

What can Elisée Reclus teach us today? How can he instruct us in geography or in the affairs of the world generally? Which, if any, of Reclus's writings can inform us or guide us in the last quarter of the twentieth century? I would suggest that a fair sampling of not only Elisée Reclus's works but of those of his allies, Peter Kropotkin and Patrick Geddes, would have a salutary effect on any geographer, young or old (compare Stoddart, 1975). A reading of their works could be a sober reminder that, despite the marvellous technological progress of the last hundred years, social progress has not proceeded apace. To be sure, there is greater equality of opportunity today and a higher material standard of living, but is there a significantly higher standard of humanity? Has there been a century's worth of progress?

Of all Reclus's interesting and provocative works, I should especially recommend two to anyone who wants to explore further – L'Homme et la terre and an article on nature sentiment published almost four decades earlier (Reclus, 1866; Reclus, 1905–1908). I should not recommend treating his works in the way that the devout pore over religious texts, but I nevertheless feel that Elisée Reclus could still serve as an inspiration to all who aspire to produce a morally uplifting as well as scientifically sound geography. His nephew Elie Faure, a well known art critic, called him a 'beacon' (Faure, 1964, 767), and he can certainly serve as a beacon and a guide to us all.

Note

For full documentation the reader can refer to my recent biography of Reclus (Dunbar, 1978). I should like to acknowledge the support of the Research Committee of the Academic Senate of the University of California, Los Angeles. An earlier version of this paper was read by Professor Paul Avrich of Queens College, City University of New York, but he is not responsible for any errors that remain.

References

Drache, H. M. 1964: *The day of the bonanza: a history of farming in the Red River Valley of the North* (Fargo, North Dakota).
Dunbar, G. S. 1974: Elisée Reclus and the great globe. *Scottish Geographical Magazine* 90, 57–66.
Dunbar, G. S. 1977: Some early occurrences of the term 'social geography'. *Scottish Geographical Magazine* 93, 15–20.
Dunbar, G. S. 1978: *Elisée Reclus, historian of nature* (Hamden, Connecticut).
Faure, E. 1964: *Oeuvres complètes*, vol. 3, *Oeuvres diverses* (Paris).
Gaidoz, H. 1875: Review of the first two livraisons ofReclus's *Nouvelle géographie universelle*, vol. 1. in *Revue critique d'histoire et de littérature* 9, 361–3.
Gaidoz, H. 1876: Bulletin géographique. *La Revue politique et littéraire. (Revue des cours littéraires*, 2nd series) 6, 35–9.
Giblin, B. 1971: *Elisée Reclus. Pour une géographie.* Thesis, University of Paris-Vincennes.
Giblin, B. 1976: Elisée Reclus: géographie, anarchisme. *Hérodote* no. 2, 30–49.
Maitron, J. 1951: *Histoire du mouvement anarchiste en France (1880–1914)* (Paris).
Meynier, A. 1969: *Histoire de la pensée géographique en France (1872–1969)* (Paris).
Moore, R. L. 1970: *European socialists and the American promised land* (New York).
Nettlau, M. 1928: *Elisée Reclus, Anarchist und Gelehrter* (Berlin).
Reclus, E. 1866: Du sentiment de la nature dans les sociétés modernes. *Revue des Deux Mondes* 63, 352–81.
Reclus, E. 1876: *Nouvelle géographie universelle*, vol. 1 (Paris).
Reclus, E. 1880: Ouvrier, prends la machine! Prends la terre, paysan! *Le Révolté* 1, no. 25, 1.
Reclus, E. 1883: *The earth and its inhabitants:Europe*, vol. 1 (New York).
Reclus, E. 1884: Anarchy: by an anarchist. *Contemporary Review* 45, 627–41.
Reclus, E. 1887: Les produits de l'industrie. *Le Révolté* 8, nos. 45, 47 and 49.
Reclus, E. 1900: Les colonies anarchistes. *Les Temps Nouveaux* 6, no. 11, 1–2
Reclus, E. 1905–1908: *L'Homme et la terre*, 6 vols. (Paris).
Reclus, P. 1964: *Les frères Elie et Elisée Reclus; ou, du protestantisme à l'anarchisme* (Paris).
Stoddart, D. R. 1975: Kropotkin, Reclus, and 'relevant' geography. *Area* 7, 188–90.
The Times, 1877, 29 December, 3. 1879, 17 December, 11. 1883, 15 January, 4. Anonymous reviews of Reclus's works, presumably all by J. S. Keltie (London).
Woodcock, G. and Avakumović, I. 1950: *The anarchist prince: a biographical study of Peter Kropotkin* (London and New York).

10

Alfred Weber and Location Theory

Derek Gregory

INTRODUCTION

In Europe the contemporary regional sciences are for the most part still predicated on various versions of positivism, and this poses acute problems of historical interpretation. In France, for example, the spatial dimension had been an insistent element of studies in political economy since the end of the seventeenth century, and it continued to have a place there until well after the turn of the eighteenth. By the early 1800s, however, its importance had waned considerably: so much so, in fact, that Dockès can claim that 'the abundant flow of ideas integrating spatial analysis and economic analysis disappears almost completely in the nineteenth century like a river vanishing in the desert' (Dockès 1969, 427–8). The conjuncture between the eclipse of the spatial tradition and the rise of a Ricardian political economy is clear enough (Scott, 1976), but what is more important here is that it also coincided with the rise of Comtean positivism. In France, therefore, a lacuna appeared between the early growth of the regional sciences and the codification of the epistemology to which, in a modified form, they subsequently subscribed (Broc, 1976). In Germany, widely regarded as the traditional home of location theory, this became a disjuncture.[1] There, the regional sciences developed throughout the nineteenth and early twentieth centuries in an intellectual milieu which was strongly resistant to positivism and its naturalist pretensions. Such a peripheral, almost cloistered existence for the regional sciences (and one which was repeated in several countries) must undermine any attempt to treat their structure and transformation in terms of cohesive paradigms or épistémés, as Kuhn (1970) or the early Foucault

(1966) would suggest.[2] A finer mesh of concepts is necessary in order to delineate the contours of specific projects with rather more precision than these sorts of formulation allow. In particular, it is of the utmost importance to discover the way in which one set of constructs was mediated by another (for example, the historical relations between the concepts of location theory, political economy and classical sociology) and to clarify the way in which these determinate discourses were bound to what we might think of as their constitutive social formations (for example, the historical relations between the ideology of location theory and the imperatives of industrial capitalism). But it is also vitally important to recognize that more is at stake than just setting the historical record straight.

I have argued elsewhere that a necessary moment in the construction of a critical science is the examination of discourse (Gregory, 1978), and that this entails an interrogation of:

(1) the *process* through which the discourse is *produced*: that is, the grounding which, in some sense, and however partially, establishes the questions its system of concepts can confront and those on which it is more or less bound to remain silent; and

(2) the *concepts* through which the discourse is *conducted*: that is, their status, dependence and logical coherence.

Massey has argued for a historical critique of location theory for reasons which correspond, very broadly, to each of these concerns:

> firstly, to set the theories in their historical context, and thereby to illustrate both their reactive nature and the role they play in that context; and secondly, to see that the approach to industrial location – that is, the nature of the theories themselves – should also take industrial location (i.e. its object) in its historical perspective (Massey, 1973).

Both of these arguments assume considerable importance in the essay which follows, where I attempt, firstly, to tie Alfred Weber's project to the social formations of late nineteenth- and early twentieth-century Germany and, secondly, to disclose a discontinuity[3] between the search for universal principles in his *Reine Théorie des Standorts*, the 'pure' theory of location which was published in 1909, and the historical specificity of his *Kultursoziologie*, the 'cultural sociology' which he initiated in 1914–15 and formalized in the decades surrounding World War II. This theoretical transition is not without its practical consequences, of course, and the examination of discourse must also entail an interrogation of:

(3) the *events* through which the discourse is *effective*: that is, the valida-
tion and legitimation of its concepts in and through the conduct of
practical life.

In so far as it is possible to use these three propositions to fashion a
(preliminary) critique of location theory, then it seems reasonably clear
that an examination of Weber's discourse is scarcely an antiquarian
exercise and that, instead, it forms part of the construction of a genuinely
human geography (Gregory, 1978, 172).

This is a bold aim and it is hardly surprising that it needs to be qualified.
In its present form the procedures on which it relies assume the existence
not only of an *internal* structure to discourse but also of a series of
extra-discursive relationships between theory and society. This is no
doubt a commonplace, but its implications are all too easily forgotten[4].
The point is this. Our ability to recognize these *external* connections (and
hence, eventually, to restructure them) is strategically dependent upon
an adequate mode of theorizing them; yet this must in turn share in and
be constituted by precisely those extra discursive relationships which it
claims to identify. This dilemma is the starting-point of modern critical
theory. Although this is obviously not the place to spell out its theses in
any detail (see Bernstein, 1976; McCarthy, 1978; Scott, 1978; Gregory,
1978; Lewis and Melville, 1978), we must at least admit before we start
that in situating Weber's project in its historical context we are inevitably
impressing our own interests on the past – an 'innocent' history is impossible – and that we run the risk of compromising our critical intentions
if we lose sight of the connections which bind these theoretical structures
to the social formations of our own times.

INTELLECTUAL TRADITIONS AND INDUSTRIAL
TRANSFORMATIONS

In the wake of the French Revolution the bond forged between reason
and nature by the Enlightenment *philosophes* was rudely broken and
replaced by a sharp polarity which set the one against the other. To
generalize in this way is to invite caricature rather than characterization,
but in a very general and schematic way it is possible to distinguish:

(1) a *rationalist* pole, which viewed history as the gradual mastery of man
over nature. From this perspective progress was synonymous with the
instrumental rationality secured by the methods of the natural sciences
and was critically dependent on their extension into the domain of the
human sciences. Man's mission was supposed to be to conquer nature
without *and* nature within, to triumph over 'the blind emotions, base-

less fears and dark superstitions [which] testified to his original state'
(Stedman Jones, 1977, 20);

(2) a *romantic* pole, which viewed history as the progressive unfolding of
an essential harmony between man and nature. From this perspective
the march of reason was leading man away from his true being,
because 'the real affinity between man and nature is not that of a being
governed by the same soulless laws of science; it is, on the contrary, a
spiritual affinity with the rhythms and patterns of nature' (Stedman
Jones, 1977, 21). This amounted to a tacit if imperfect recognition that,
as Marcuse was to say later, 'the scientific method led to the ever-
more-effective domination of nature [and] thus came to provide the
pure concepts as well as the instruments for the ever-more-effective
domination of man by man through the domination of nature' (Marcuse, 1972, 130).

One of the enduring tragedies of European thought in the nineteenth
century was its failure to accommodate the necessary tension between
these two, and it was not until the twentieth century that German philosophy came to resist the universal and exclusive claims made by what
Habermas has described as the empirical-analytic sciences constituted by
a *technical* interest and the historical-hermeneutic sciences constituted by
a *practical* interest. In Habermas's view these two cognitive interests,
which surfaced through the rationalist and romantic poles, are submerged in two primary levels of social action which constitute two distinct forms of knowledge by determining their respective objects of study
and the criteria for making valid statements about them. The first of these
is *production:* the labour process has to achieve technical control over its
materials and hence necessitates a conception of science which can
provide intersubjectively agreed (nominally successful) predictions about
them. But this is not self-sufficient, says Habermas, because the process
of negotiation on which it relies is an intrinsically hermeneutic one: only
through an act of *communication,* the second primary level of social action,
can agreement be reached over these procedures and their consequences.
This necessitates another conception of science which is assigned the task
of interpretation, of sustaining the reciprocity of meaning on which
practical life depends (Gregory, 1978, 68–9). As Giddens has shown, it is
by no means as easy to differentiate production and communication as
Habermas would want, but it *is* the case that, whatever the shortcomings
of his characterization, the insistent rejection of the 'claim to universality'
of the empirical-analytic and the historical-hermeneutic sciences remains
a persuasive one (Giddens, 1977).'
But hindsight is the privilege of the present, and throughout the nineteenth century the polarity between the two increased; while France

declared its commitment to rationalism, Germany allied itself to romanticism. This is to speak too simply, of course (Hawthorn, 1976, gives a
finer set of distinctions; see also Simon, 1963): the attachment was a
precarious one, and although in Germany it was elaborated in many
different ways these never coalesced into a single, coherent system. As a
result it never achieved much more than a very general status which
informed intellectual endeavour but which failed to *determine* its precise
form. What was more, the privileges accorded to the so-called *Geisteswissenschaften* and their disclosure of the various expressions of man's
inner being had a distinctly hard edge to them. It was true that at the
beginning of the century Kant, Schelling, Schleiermacher, Fichte and von
Humboldt had issued a series of programmatic memoranda which advocated a 'pure' idea of academic freedom. Theirs was a spirited renuncia-
tion of the technical rationalism which had dominated the University of
Halle, the locus of what was then called cameralistic education (the
science of administration and statecraft), and a call for a return to the
classical scholasticism which revered learning for its own sake. Even in
very narrow terms, however, these attempts to free the universities from
direct state control were at best partially successful, and although they no
longer bowed to its immediate and specifically technical demands their
pupils continued to file behind the high desks of the Prussian bureaucracy.

Even so, what Hawthorn (1976, 141) describes as this 'tacit alliance'
managed to win for the universities a licence which at least ensured that
in the years ahead their historical-hermeneutic concerns would be respected and that the illumination of a distinctively German *Geist* would
move steadily towards the centre of the academic stage. But in a much
deeper sense this spiritual mission resonated with profoundly political
implications, inasmuch as its evident debt to Hegel reaffirmed the idealist
conception of the state 'as not merely the arbiter and administrator of
society but as the institution which defined its ends and so transcended it'
(Hawthorn, 1976). With the establishment of the *Reich* in 1871, therefore,
the enlistment of the universities in the cause of *Bildung* and *Kultur*, in the
cultivation of man's inner being and the contemplation of his spititual
achievements, came to be a task of national importance which recalled the
doctrine of the *Kulturstaat*, the 'cultural state' which had emerged in the
absence of unification at the turn of the century, but which at the same
time went way beyond it. The prominence and sense of purpose of
intellectuals, at once cultural and political, prompted Ringer to describe
theirs as a 'mandarin ideology', and this is an entirely convincing image
(Ringer, 1969).

But Bismarck's priorities were not restricted to the cultural-political
sphere, and much of his grand design was cut from an economic template.
By this time the industrial leadership of Europe was passing from Britain to

Germany. In the decades before 1871 Germany had gradually turned
from one of the main export markets for British manufacturers into a
self-sufficient industrial economy and after 1871 German manufacturers
began to make rapid inroads abroad. The statistics set out in figures 10.1
and 10.2 are only indications of the change, and little more than approxi-
mations at that, but they speak for themselves. In some respects, perhaps, they shout too loudly. Germany still had to challenge Britain's
commercial hegemony and, even within the industrial sector, many labour
processes retained elements of domestic or *Handwerk* organization.
Further, as the tremors on both graphs suggest, the years between c.1873
and c.1895 were punctuated by a series of economic crises which together
added up to a major recession, marked by tumbling prices, falling
production and rising unemployment. But this in fact accelerated the
concentration of capital:

> many industrialists agreed that in this crisis situation those in big
> industry would have to stand together to survive. The formation of
> cartels seemed to be one of the more promising means of riding out
> the storm . . . There are clear indications of informal cartels in the
> period before the Depression, but the economic crisis speeded up
> the tendency for a 'protective collectivism'. (Kitchen, 1978, 154)

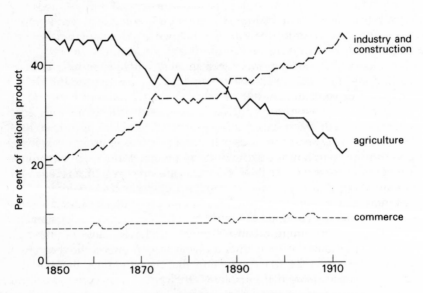

FIGURE 10.1   Economic change in Germany c. 1850-c.1913
(Source: B. Mitchell, *European Historical Statistics*, 1975, Table K2)

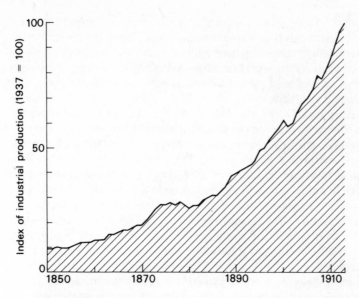

FIGURE 10.2   Industrial growth in Germany c. 1850-c.1913
(Source: B. Mitchell, *European Historical Statistics*, 1975, Table E1)

Whatever qualifications might be made, therefore, it was evident that, in Landes's arresting phrase, the industrial transformation of and even the consequences of the economic crisis in the new Germany 'stuck in John Bull's craw' (Landes, 1970); and what was also immediately obvious was that it deeply troubled the German academic community.

Writing in 1904, only five years before Weber published his industrial location theory, Eucken described 'a sinking of life into the profane, the secular, the vulgar' in the middle of 'a remarkable progress at the periphery of life'. This had been achieved through the development of the empirical-analytic sciences, and as a (paradoxical) result of the technical control which they had established 'work emancipated itself from men; it formed huge complexes, which increasingly generated their own forces and followed their own laws. Thus there arose a sharp conflict between work and soul' (Eucken, 1904, 4, 108). Anxieties like this could be multiplied many times over. To be sure, the trauma – even the decadence – of the *fin de siècle* was not confined to Germany, but it assumed a peculiar acute and agonizing form there because German academics believed that they had singularly failed in their duty as what Jaspers called the 'bearers of tradition' (Jaspers, 1923, 61). All the time that they identified that tradition with (and hence defined their role as) the *exclusive* preservation of the practical interest of the historical–hermeneutic sciences they were, of course, bound to do so, since the labour process requires technical control over the forces of production and thus necessitates a conception of science which can provide successful predictions about its materials (and, for that matter, about the men and women who use them). This is all the more obvious in a period of industrial expansion, and a degree of compromise was already an established part of academic practice, if not always of its rhetoric.

Thus in 1867 Dilthey had accepted that the human sciences would in some measure, have to engage the material needs of society; but, he continued, 'this professional education will fit the industrial for the higher functions only in proportion as it goes beyond the limits of a technical training' (Hawthorn, 1976, 143). This insistence on the importance of life as it was experienced outside the narrowly technical domain was scarcely a novel one, but it now assumed a new urgency. In 1872 the *Verein für Sozialpolitik* (Association for Social Policy) was founded, to inject the all-important moral dimension into economic life. But the commitment, however strident it had become, was no more precise than it had been before, and the *Verein* was racked by the disputes of the famous *Methodenstreit* right down to World War I (Hawthorn, 1976; Frisby, 1976). By the early years of the twentieth century German academics, disunited and bewildered, were coming to realize that the specific forms being assumed by the empirical–analytic sciences conveived of and reproduced their conditions of existence in terms which threatened to undermine the conditions of existence of other conceptions of science. It was no accident, therefore, that their sense of despair was as profound as it was protracted. Even the emergence of a revivified *Lebensphilosophie*, which owed much to Dilthey, could not arrest the continued universalization of the technical interest which marked the consolidation of industrial capitalism in Western Europe, and the *Krise der Wissenschaft* continued through the 1920s into the 1930s.

ALFRED WEBER AND THE CULTURAL–INTELLECTUAL
CRISIS

It was against this turbulent background that Alfred Weber embarked on his academic career. The son of a magistrate, he was born in July 1868 at Erfurt, a small provincial town in the heart of Germany. The following year his father (Max senior) entered the *Landtag*, and soon afterwards the *Reichstag*, as a National Liberal deputy. This prompted the family to move to Charlottenburg, a suburb on the outskirts of Berlin which was to remain their home for many years to come. It was there that Alfred and his brother Max, four years older than himself, were first introduced to 'the special character of political life' (Marianne Weber, 1975, 40). But however insistent the external excitements of politics may have been to his elder brother – and Max's wife Marianne recalled that

the knowledge of world history in the making that young Max
acquired in this manner remained vividly in his mind for forty years

– and however much these sort of arguments were to mean to Alfred later
in his life, his thoughts were then resolutely introspective. His mother
Helene, an unrelenting evangelist, confided that

> he is having a very hard time. This I notice by the vehemence and
> stubbornness with which he seizes upon every opportunity to prove
> that every other view has at least the same justification and proba-
> bility as the Christian religion. Then he cites Strauss's *Das Leben
> Jesu* and the philosophy of Kant. I stand there and feel sorrowful and
> sad that I cannot help him, because I lack the right words at the right
> time. (Marianne Weber, 1975, 59).

He was always very close to his mother, but for all her concern his doubts
persisted. Just before his eighteenth birthday he wrote several long letters
to Max, then away on military service in Strasbourg, asking about his
forthcoming confirmation and, according to Marianne, 'he gladly took
the advice of his admired and precocious brother' (p. 98).

The relationship between the two was neither as constant nor as
harmonious as this anecdote implies, and Marianne's memoir systematically evades the conflict which grew up between the brothers as Alfred
gradually came to resent being overshadowed by *der grosse Max*. Even so,
when Alfred went to university in Berlin he followed closely in his
brother's footsteps, reading the history of art, jurisprudence and, finally,
political economy and social science. In 1897 he submitted his doctoral
thesis on the sweating system in the clothing industry, research which he
had completed under the watchful eye of Gustav Schmoller, and he
published two extracts from it in the *Archiv für Sozialwissenschaft und
Sozialpolitik*, a series edited by Max, and a general statement on the
political economy of workshop industry in *Schmollers Jahrbuch* for 1902.
He remained at Berlin from 1899 to 1904 as a *Privatdozent*, and there he met
and fell in love with Else von Richthofen. She was a strikingly beautiful
woman, the daughter of Baron von Richthofen, and in Berlin to complete
her own doctoral thesis under the supervision of Max: the brothers were
constant rivals for her affections, and this provided yet another bone of
contention between them[5].

In 1904 Alfred moved to Prague to take the chair of political economy at
Karls-Universität, where he became a close friend of Max Brod and,
through him, received an introduction to the ideas of *Lebensphilosophie*
which were to stay with him, in one form or other, for the rest of his life.
(In fact, Brod portrayed Weber as Professor Westertag in his novel *Jugend
im Nebel*, and compared him so favourably with Dehmel and Schopen-

hauer that his lectures became extremely fashionable in Prague's literary circles and *salons*) (Marianne Weber, 1975, 226). It was in Prague, too, that Weber met George Pick, who later wrote the mathematical appendix for his location theory. Three years later he was offered the chair of political economy and social science at Heidelberg. Marianne records that 'he found it hard to part from Austria and hesitated before making a decision', a decision which, one suspects, must have been complicated by the attraction of Else, who had by then married a Heidelberg political economist, Edgar Jaffe, and above all by the intervention of Max, who had recommended Georg Simmel for the post. But in the end Alfred accepted, not without misgivings, and, Marianne continues, Max was 'delighted; the brothers had once being very close to each other' (Marianne Weber, 1975, 366; Green, 1974, 226).

The move to Heidelberg was decisive: it remained Alfred Weber's home until his death in 1958, and its intellectual circles were largely responsible for rounding out his commitment to the principles of *Lebensphilosophie*. He was now ready to recognize the importance of a cultural revival, and the sense of a spiritual crisis had begun to affect him deeply. But he was, inevitably, more circumspect that he had been as an adolescent, and he wanted to know first of all whether it was 'sensible for us to argue about cultural and social motives when perchance we are simply fettered by the iron chains of hard economic forces?' Nobody could miss what Eucken had described as the centrifugal and centripetal effects of industrial urbanism, the rise of the great cities (figure 10.3)[6], but while it was all very well to talk of the moral debilitation of *Asphaltkultur*, he doubted whether it was 'possible for us to arrive at any conclusion about its causes when we do not possess as yet any real knowledge of the general rules determining the location of economic processes – when the purely economic laws which without any doubt somehow influence location are not discovered' (A. Weber, 1909, 2–3). His early lectures at Heidelberg, which were attended by the young Walter Christaller (Christaller, 1972, 602), were therefore designed to clarify the ways in which, as he was to say later, 'the locations of the industries form the 'substance' (I do not say the cause) of the large agglomerations of people today.'[7]

To this extent, his location theory represented a certain qualification of the anti-naturalism which dominated contemporary discourse. Isard has gone further, to suggest that Weber's attempt to develop 'a theory of the transformation of locational structures' was 'undoubtedly influenced by the writings of Roscher and Schäffle', both of whom 'were primarily concerned with discovering whether or not there are any natural laws or regularities in the evolving locational structure of economies'.[8] Isard interprets this in such a way that Weber is placed firmly in the positivist camp, thereby providing an historical warrant for the contemporary prosecution of the regional sciences; but this, I think, is substantially

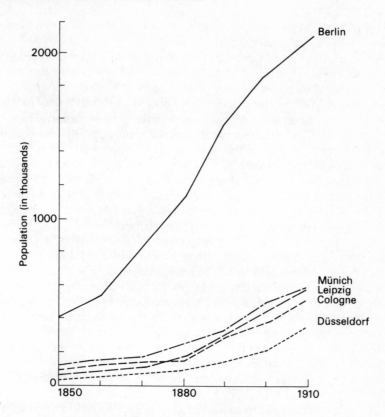

FIGURE 10.3   Urban growth in Germany c.1850–c.1910

(Source: B. Mitchell, *European Historical Statistics*, 1975, Table B4)

incorrect. Weber's training in political economy had been with the so-called Younger Historical School of Schmoller rather than the Older Historical School of Roscher. What was more, he was certainly aware of Max's contributions to the *Methodenstreit* – they were both leading members of the *Verein* – which had included a swingeing attack on Roscher, published in *Schmollers Jahrbuch* for 1904. In his critique Max had conceded that the discovery of empirical regularities 'may have extraordinary *heuristic* value', but he was not prepared to accept this as 'the ultimate *goal* of any science': he insisted that these generalizations 'had no causal status'. Instead (and at best) 'correlations of this sort could only constitute the subject matter of a scientific investigation, an investigation which would begin only after these correlations had been established.' In Max's view the propositions of the social sciences could only achieve a legitimate causal status by referring, in some fashion, to the meanings and intentions of the actors involved. Hence an adequate social science would have to be historically specific (M. Weber, 1975).

There was, of course, a good deal more to his essay than this, but even in this abbreviated form it helps to account for the distinction which Alfred drew between 'substance' and 'cause' and, more important, for the explicit restrictions he placed on his abstract formulation of location theory. He insisted on the viability of a mathematical presentation of its arguments – the recourse to mathematics was not a mere *Spielerei*, he said, but a clear necessity – but at the same time he accepted that it was an incomplete solution. In particular, a wholly abstract formulation would 'hang in the air' (*in die Luft stösst*): eventually, it would have to engage specific forms and structures if its practical significance were to be realized. Moreover, he made it perfectly clear that 'the kind of industrial location we have today is not entirely explained by the 'pure' rules of location and therefore is not purely 'economic'. It results to a large extent rather from very definite central aspects of modern capitalism and is a function of modern capitalism, which might disappear with it. It results, we may say in hinting at the main point, from degrading labour to a commodity bought today and sold tomorrow, and from the ensuing laws determining the labour market and from the local agglomeration of workers created thereby.' 'I do not wish to assert,' he concluded, 'that the creation and development of labour location can be explained by economic reasons; but if it can be so explained the reasons will be related to the position which the particular economic system gives to labour.' (A. Weber, 1909, iii–iv, 12–13, 225).

In these programmatic remarks Weber is grappling with two problems of articulation posed at two levels of discourse: in *epistemological* terms he confronts the relation between the abstract and the specific and in *ontological* terms he confronts the relation between the material and the cultural. In one sense, therefore, his project can be seen to parallel that of his brother, and there are obvious affinities between Alfred's interest in agglomeration and Max's concern with bureaucratization.[9]

But soon after he settled in at Heidelberg Alfred joined the Georg *Kreis*, a tightly knit circle of anti-modernist poets and social critics formed around Stefan Georg, and which included 'a remarkable number of the liveliest minds in early twentieth–century Germany' (Mitzman, 1970, 262–3). Max attended many of their meetings as well, but although he apparently enjoyed the debates and discussions he 'was treated as an honourable enemy by the poet's followers' (Green, 1974, 163; Marianne Weber, 1975, 455–64). In fact, he made no secret of his pessimism about the *Kreis's* attempts to serve as the catalyst for a new era of high culture and his belief that they were 'doomed to disintegrate on the iron rock of real social and economic relations' (Mitzman, 1970, 265). But this did not stand in the way of Alfred, who by then had seen the need to break away from his earlier ideas and reformulate them not only in a wider context but in an entirely different framework. His original intention had been to complement the abstract theory of location by a specific theory of location under capitalism, but all he ever managed was the simple outline sketch contained in a brief essay in the *Grundriss der Sozialökonomik*, with its facile and essentially empiricist distinction between *der gebundene Kapitalismus* and *der freie Kapitalismus*. [10] But he had come to regret these early studies (J. Barnbrock, personal communication), and in the years which followed, interrupted only by a brief (and in many respects a wounding) excursion into national politics, [11] he devoted himself to the formation of a cultural sociology. This was not only more in accord with the demands which he had imperfectly recognized in his location theory, and which, one can argue, had been stifled by the neo-Kantian problematic in which they had been formulated, but also in sympathy with the idealist aspirations of the celebrated *Heidelberger Geist*, which was then seeking, in Green's phrase, 'new depths of meaning'. This led Weber to the gradual embrace of neo-Hegelianism. Unlike previous attempts to find in Hegel a legitimation of a doctrine of *state* right and authority the new philosophical order was committed to an exploration of the richness of *man's* unique spirit. This ruled out any easy separation of public and private responsibilities – and like many academics Weber became a vocal member of the pre-war pacifist movement (Chickering, 1975) – and it demanded a redefinition of existing concepts of culture.

In 1915, only a year after completing his entry on location theory for the *Grundriss*, Weber published some preliminary reflections on the nature and method of cultural sociology (Weber, 1915), and once the war was over and his involvement with the ill-fated Democratic Party had come to an end he elaborated these ideas into a set of formal principles (Weber, 1920) which he used in a study of ancient Egypt and Babylon (Weber, 1926) His efforts culminated in the monumental *Kulturgeschichte als Kultursoziologie*, which appeared in 1935 and was dedicated to Else. The fundamental feature of Weber's work throughout this period was the attempt to establish what Aron treats as 'a scientific theory of history' (Aron, 1957, 44), based on a distinction between three inter-related spheres of historical reality. These were:

(1) *Gesellschaftsprozess* ('societal process'): the unchangeable – but not unchanging – material realities of the world; these were held to be 'incontrovertible, so that we can only respond to them, take them into account in our behaviour' (Neumann, 1948), which we do through:
(2) *Zivilisationsprozess* ('civilization process'): the intellectual advances which man makes through the instrumental rationalism of the natural sciences. In so far as their methods appeal exclusively to reason they are supposed to be intrinsically communicable and hence (in principle)

capable of universalization. Weber thus saw 'this human mastery over nature [as] an irreversible process which will finally lead to a unitary civilization and generally accepted ideas' (Neumann, 1948). But at the same time he insisted that the social sciences had been wrong to suppose that this was the only dimension of progress and that history consisted solely in the dialectic between *Gesellschaftsprozess* and *Zivilisationsprozess*, so that the non-material was somehow simply derivative, even epiphenomenal: 'a mere shadow cast by the wheels of an onrushing locomotive.' What such a vision failed to take into account was:

(3) *Kulturbewegung* ('culture movement'): the unique and, in a particular sense, non-rational, hence non-communicable imperatives which admit of no universal laws because they are governed by the soul rather than the mind.

Weber tied each of these spheres to a distinctive and deep-seated 'will': *Gesellschaftsprozess* to the natural-vital will, *Zivilisationsprozess* to the rational-utilitarian will and *Kulturbewegung* to the ideal-transcendental will. [13]

For all its faults, this scheme clearly reaches back to the very problems of articulation with which Weber had stuggled in his location theory, while at the same time it reaches forward to an uneasy accommodation of the rationalist/romantic polarity. What makes it an unsatisfactory compromise is the absence of theoretical determination: 'none of these spheres is the cause of historical development, and none is determined solely by an immanent law. The development as a whole can only be understood in terms of the combined action of these three spheres and their reciprocal interaction' (Aron, 1957, 48). If this is the case, and if these spheres are meant to be something more than very general principles of explanation, then it is necessary to show how, precisely, they are constituted: without doing so, they must remain as the emptiest of propositions. Tying them to *wills* is no solution because, as Habermas has realized of his own attempt to specify *cognitive interests*, it still remains to show that these sort of categories – and the two are by no means equivalent, of course – are the result of more than an arbitrary impulse, 'the product of an embarrassment which points to more problems than it solves' (Habermas, 1974, 14). Idealism, I suggest, is unable to confront problems of this order, far less resolve them. But Weber's project is significant, nevertheless, because of its unequivocal affirmation of the importance of historical specificity and, connected to this, of the belief that 'the progress of science is in no way a guarantee of spiritual growth', that in a vitally important sense 'the triumph of civilization marks a return to barbarism' (Aron, 1957).

Weber was not alone in these views, and much of his cultural sociology can be related really very explicitly to the writings of Dilthey, Scheler and, in particular, Husserl. Husserl's Krisis lectures, which he delivered in Prague in 1933, declared the European nations to be 'in a critical condition' which would end in either 'the ruin of Europe alienated from its rational sense of life, fallen into barbarian hatred of spirit; or in the birth of Europe from the spirit of philosophy, through a heroism of reason that will definitely overcome naturalism' (Husserl, 1970). This was the object of his transcendental phenomenology, nothing less than the restoration of the spirit to itself through an act of pure, inner contemplation, and Husserl clearly believed that opposition to naturalism and opposition to fascism had to be all of a piece. Both reduced man to an object in an objectified world. Although Husserl died before war broke out, he left behind a remarkably rich set of arguments which together stand as a powerful testimony to his vision of the times (Pivcević, 1970).

Weber's own formulations were much less striking, and the development of his cultural sociology was cut short by the war. But like Husserl he was an implacable opponent of fascism, and in 1933, the year of the so-called *Volkwerdung*[14], he categorically refused to fly the Nazi standard from the roof of his Institute. This was a dangerous decision, and he was violently abused and assaulted by the Hitler Youth and the student Nazi movement. Throughout the official interrogations which followed he stood his ground, and he eventually resigned rather than serve under the Third *Reich*. He spent the war years as a private citizen in Heidelberg, an isolated but, as MacRae rightly says, an uncompromised scholar (see D. MacRae in Aron, 1957, vii). In the closing years of the war, and in the immediate post-war period during his service with the Allied Control Commission (E. A. G. Robinson, personal communication), he was able to return to cultural sociology, and in doing so he came much closer to Husserl's conception of the transcendental task.

He wanted to speak, he said, to those who 'roused into awareness, are willing and able to translate their outward experience into an inner one, an experience that plumbs the depth of their souls, which always – and this is the direction I stand for – means transcendental experience' (Weber, 1946, 155). The purpose of his remarks was to try to formulate a way of thinking about the enormities and atrocities of the war which, while it did not shirk from individual responsibilities, would at the same time offer some hope for the future. He was sure that the answer was not to be found in the narrowly psychological domain: 'we shall never grasp the real nature of the process like this' (Weber, 1946, 159–161). Instead, the actions executed in the name of the Third *Reich* had to be seen as 'the outbreak of forces, certainly stimulable by psychology and capable of being roused under certain [material] conditions, but forces coming from greater depths than psychology ever plumbed. It was a sudden darkening of mind that set in, an occultation in which one felt the uncanny wing-beat . . . of the *dark-daemonic* forces: there is no other term for their supra-personal and at once transcendent power'. This was *not* apologia masquerading as rhetoric: Weber insisted that 'we have suffered all this knowingly and with open eyes'; but, he continued, that being so, 'shall we then be so shallow-minded as to ignore the *deeper* level that is apparent here, the transcendental and metaphysical level known to the men of earlier times, of which but *one* aspect, one of its many sides, has, under our very eyes, made itself master of our lives for a spell?'

The crux of his argument was that the working class had been 'brutally expropriated' of precisely these 'intimations of transcendence', 'systematically robbed of every higher view of life'. What was therefore needed, he believed, was an *intellectual* elite to 'act as continual models' for the rest of society. But if this was a return to his pre-war credo, it was one with a difference.

> We, the intelligentsia, want freedom, intellectual freedom. We know well enough that without it our initiative is broken, we are helplessly fluttering birds caught in a cage, picking up bits of food and little else, strangers to ourselves. But can there be intellectual freedom except on the basis of political freedom? And is political freedom today, when the confused, traditionless, materialistic masses are awake and jealous of their rights and eager to assert them, is political freedom possible except as the political freedom of the masses, that is, their self-government? Of course not (Weber, 1946, 166).

Weber maintained that the solution to all this, and one which was credible in both intellectual and political terms, was what he called Free Socialism.

Through its Action Group in Heidelberg the new movement proclaimed that 'the new historical situation was no longer the class struggle but the preservation of freedom in the (capitalist or socialist) 'monopolist, totalitarian and bureaucratic state' (A. Weber and Mitscherlich, 1946; for a critical commentary see Graf, 1976). This can also, I think, be seen as a logical extension of Weber's early studies, transformed by his subsequent encounter with idealism; but it proved to be as intellectually and politically bankrupt as the Democratic Party had been before. Hawthorn regards it as 'a curious mixture of liberalism and syndicalism, more directly reminiscent of Tönnies than of anyone else' (Hawthorn, 1976, 184), a fair enough description, and even on its own terms it could be seen to be seriously inadequate since it said nothing about how free socialism would manage to combine what Weber had previously identified as 'the dictates of civilization and the promise of culture' (Hawthorn, 1976). It is not altogether surprising, then, that Weber's nomination by the Communist Party for the Presidency of West Germany should have been little more than a cynical joke, designed to draw attention to what was seen as the charade of supposedly free elections in an occupied and divided state. In any event, it was certainly not intended as an endorsement of his programme, and it came to nothing.

CONCLUSION

All sorts of conclusions can no doubt be drawn from what I have said so far, but I want to underline just two.

In the first place, Weber's location theory has long been the object of criticism. Many writers have argued that the primacy it accords to cost-minimization is untenable, and that a more credible rationale would be profit-maximization; others have suggested that transport costs are no longer of such critical importance as Weber once supposed; still others have objected to his assumptions of perfect competition; and, finally, several have argued for a general location theory capable of addressing more than one sector of the space-economy. These criticisms are, of course, well known; but however telling they might be they are all made from *within* the established problematic of location theory. Barnbrock has suggested that although the German schools of historical economics had but a marginal impact on the development of location theory 'the manner in which they embedded the social life process in the organic process of social development in a particular nation state' was a potentially impor-tant contribution. 'In addition, they emphasized that each locational process has to be seen as part of a complex socio-cultural phenomenon; (Barnbrock, 1976, 132). But, as I have just said, these points were received in silence; yet they were precisely what Weber had to offer, and whether we follow his particular trajectory or not (which is, in the present context, a side issue), we must at least concede that in coming to terms with the consequences of them Weber was obliged to move well beyond what Massey (1973, 38) has correctly identified as the 'confusion implicit in attempting to conceive of a completely autonomous location theory'. In other words, I suggest that the momentum of any critique of location theory – its efficacy – is to be judged in large measure by the distance it manages to move *outside* the traditional domain of such a theory.

In the second place, and connected to this, it ought to be clear that Weber's whole work was deeply imbricated in the ideological and material imperatives of German society. But, more than this, Weber was responding to a series of critical dilemmas which made him keenly and constantly aware of the practical implications of his theoretical formula-tions. Theory and practice were necessarily conjoined. Whether his solu-tions were acceptable is, again, beside the point: I for one do not find them so, and I would certainly wish to resist his slide to elitism. But, neverthe-less, Weber always showed himself unwilling to accept the narrowly technical conception of committed scholarship which has dominated so many of the contemporary regional sciences. His was much more than an empty appeal to 'social concern': it was instead a clear recognition that intellectual inquiry is part of – and ultimately responsible to – the conduct of practical life.

Acknowledgments

I am particularly grateful to Mark Billinge, Jörn Barnbrock, Anne Buttimer, Paul Claval, Nick Entrikin, T. W. Freeman and Roger Lee for their comments. I was able to consult the first German edition of Weber's location theory through the courtesy of the Library of the Vereins Deutscher Eisenhüttenleute in Düsseldorf.

Notes

[1] See, for example, Isard, 1956, 27:

> the reaction of German thought to classical teachings, which precipitated the rise of the German historical school, ploughed the ground for contributions in the field of *Raumwirtschaft*. In the study of the stages of economic development, the spatial structure of economic processes was necessarily a primary concern. And, with the impress of the Lausanne school of thought upon German economics, it was almost inevitable that attempts would be made at a fusion of space with general equilibrium analysis.

[2] I exclude Foucault's *L'archéologie du savoir* (Paris 1969) from this verdict, and in doing so follow Lecourt, D. *Marxism and epistemology: Bachelard, Canguilhem, Foucault* (London 1975).

[3] In an early version of this essay I wrote 'epistemological break' here, but the remarks contained in B. Hindess, *Philosophy and methodology in the social sciences* (Hassocks 1977) convinced me that there were such formidable difficulties with the Bachelard–Althusser scheme from which the term derives that it was necessary to substitute the more anodyne 'discontinuity'; but cf. Barnbrock (1976), which speaks of an epistemological break between the Kantianism of Part I of *Der isolierte Staat*, published in 1826, and the Hegelianism of Part II, published in 1850.

[4] Cf. P. Saey, Marx and the students of space *L'espace géographique* 1 (1978) 15–25 fn19: 'The neglect of (science as a societal activity, i.e. an activity that takes place in a specific society) is a significant weakness which Claval shares with nearly all geographers, including myself, who wrote on the history and methodology of their discipline.' But see R. J. Johnston, *Geography and geographers: Anglo–American human geography since 1945* (London, 1979).

[5] This needs some explanation, since Max Weber had succeeded Knies to the chair of political science at Heidelberg in 1896; when he fell ill, Else visited him in Berlin, where he was enjoying a brief convalescence at the family home. She returned to Heidelberg to receive her doctorate in 1901. She was for a time a factory inspector, but this was presumably not the direct result of Alfred

Weber's interest in industrial location. I am grateful to David Stoddart for drawing the connection between Alfred and Else to my attention; see Green, 1974.

[6] In 1871 33% of the population of Germany had been classified as urban; by 1890 this had risen to 43%, by 1900 to 54% and by 1910 to 60%. These figures suggested to Clapham 'a whole nation rushing to town', and the rush was greatest into the greatest towns: in 1890 11% of the population lived in cities of over 100,000 and by 1910 this had virtually doubled to 21% (Lee, 1978).

[7] A. Weber, 1909, 6. Similarly, in the preface (which was omitted from the English translation) he claimed that the location of industry provided 'one of the keys' to contemporary urbanization and to 'a whole host of other social and cultural forces'.

[8] Isard, 1956, 17 and fn; but note that in the body of his text Weber made only one reference to Roscher and Schäffle and that in passing: A. Weber, 1909, 6. In the first *Excurs*, however, which was also omitted from the English translation, Weber did offer a more sustained (and critical) commentary on their previous contributions, and prefigured the emergence of locational analysis in human geography as such.

[9] In fact Alfred published an essay on bureaucracy in 1913 in which he calculated that Germany had 860,000 bureaucrats in industry (almost ten times as many as in 1882) to administer 8,500,000 workers: A. Weber, Die Bureaukratisierung und die gelbe Arbeiterbewegung. *Archiv für Sozialwissenschaft und Sozialpolitik* 37 (1913). His translator drew attention to the connection between agglomeration and bureaucratization in C. J. Friedrichs and T. Cole, *Responsible Government: a study of the Swiss Civil Service* (Cambridge, Mass. 1932) 28fn, but in a highly – and, in the light of his translation, a curiously – abbreviated form.

[10] 'Controlled' and 'free' capitalism: A. Weber, Industrielle Standortslehre: allgemeine und kapitalistische Theorie des Standorts. *Grundriss der Sozialökonomik* VI (Tübingen 1914) 59–86. Weber also planned an intermediate stage between the abstract and specific theories, a series of substantive studies of German industry since 1860. Eight of these volumes were published under his general editorship as *Uber den Standort der Industrien, II: die deutsche Industrie seit 1860* (Tübingen 1913–31), but he wrote none of them himself.

[11] Weber helped to found the Democratic Party and played an important part in fashioning the democratic framework of the Weimar Republic; 'but this final attempt by the liberal intellectuals to transcend the actual divisions of German politics and society through the ideal of freedom proved to be an epilogue rather than a beginning' and such popular support as it had to start with soon evaporated: L. Krieger, *The German Ideal of Freedom: history of a political tradition* (Boston 1957) 464–6. Weber was never much of a populist, and like many of his contemporaries he saw the intellectual's role in strictly elitist terms: his autocratic interventions at meetings of the Democratic Party were instrumental in his enforced resignation: see W. Struve, *Elites against Democracy: leadership ideals and bourgeois political thought in Germany, 1890–1933* (Princeton 1973).

[12] Weber described these years in Heidelberg as

intense above all else because whatever happened there intellectually took place amid a strong and productive exchange of all man's powers. It was fully absorbed in and permeated by the new life that had begun to develop in Germany since the turn of the century. It was intellectually and personally exciting, and open on all sides (Green, 1974, 140–1).

[13] For more detailed accounts of Weber's project, see Aron, 1957, 43–51, and M. Luoma, *Die drei Sphären der Geschichte* (Helsinki 1959). Few of Weber's writings have been translated into English, but one of the exceptions is *Fundamentals of culture-sociology: social process, civilizational process and culture-movement* (translated by G. H. Weltner and C. F. Hirshman, New York 1939).
[14] 'A people becoming itself': one of the many Nazi designations for their seizure of power. See W. Allen, *The Nazi Seizure of Power* (London 1965).

References

Aron, R. 1957: *German sociology* (London).
Barnbrock, J. 1976: *Ideology and location theory: a critical inquiry into the work of J. H. von Thünen.* Unpublished Ph.D. dissertation, Johns Hopkins University, Baltimore.
Bernstein, R. J. 1976: *The restructuring of social and political theory* (Oxford).
Broc, N. 1976: La pensée géographique en France au XIXe siècle: continuité ou rupture. *Revue géographique des Pyrénées et du sud-ouest* 47, 225–47.
Chickering, R. 1975: *Imperial Germany and a world without war* (Princeton).
Christaller, W. 1972: How I discovered the theory of central places: a report about the origin of central places. In English, P. and Mayfield, R., editors, *Man, space and environment: concepts in contemporary human geography* (Oxford) 601–6.
Dockès, P. 1969: *L'espace dans la pensée économique* (Paris).
Eucken, R. 1904: *Geistige Strömungen der Gegenwart* (Leipzig).
Foucault, M. 1966: *Les mots et les choses* (Paris).
Frisby, D. 1976: *The positivist dispute in German sociology* (London).
Giddens, A. 1977: Habermas's critique of hermeneutics. *Studies in Social and Political Theory* (London) 148–54.
Graf, W. D. 1976: *The German Left since 1945* (Cambridge).
Green, M. 1974: *The von Richthofen sisters: the triumphant and tragic modes of love* (London).
Gregory, D. 1978: *Ideology, science and human geography* (London).
Habermas, J. 1974: *Theory and practice* (London).
Hawthorn, G. 1976: *Enlightenment and despair: a history of sociology* (Cambridge).
Husserl, E. 1970: *The crisis of the European sciences and transcendental phenomenology* (Evanston).
Isard, W. 1956: *Location and space-economy* (Cambridge Mass.)
Jaspers, K. 1923: *Die Idee der Universität* (Berlin).
Kitchen, M. 1978: *The political economy of Germany, 1815–1914* (London).
Kuhn, T. 1970: *The structure of scientific revolutions*, second edition (Chicago).
Landes, D. 1970: *Who unbound Prometheus? Technological development in Western Europe* (Cambridge).
Lee, J. J. 1978: Aspects of urbanization and economic development in Germany 1815–1914. In Abrams, P. and Wrigley, E. A., editors, *Towns in societies: essays in economic history and historical sociology* (Cambridge) 279–93.
Lewis, J. and Melville, B. 1978: The politics of epistemology in regional science. In Batey, P., editor, *Theory and method in urban and regional analysis* (London) 82–100.
Marcuse, H. 1972: *One dimensional man* (London).
Massey, D. 1973: Towards a critique of industrial location theory. *Antipode* 5, n.3.

McCarthy, T. 1978: *The critical theory of Jürgen Habermas* (London).
Mitzman, A. 1970: *The iron cage: an historical interpretation of Max Weber* (New York).
Neumann, S. 1948: Alfred Weber's conception of a historico-cultural sociology. In Barnes, H., editor, *An introduction to the history of sociology* (Chicago).
Pivcević, E. 1970: *Husserl and phenomenology* (London).
Ringer, F. 1969: *The decline of the German mandarins: the German academic community, 1890–1933* (Cambridge, Mass.).
Scott, A. J. 1976: Land and land rent: an interpretative review of the French literature. *Progress in Geography* 9.
Scott, J. 1978: Critical social theory: an introduction and critique. *British Journal of Sociology* 29, 1–20.
Simon, W. 1963: *European positivism in the nineteenth century* (Cornell).
Stedman Jones, G. 1977: The Marxism of the early Lukács. In New Left Review, *Western Marxism: a critical reader* (London). 11–60.
Weber, A. 1909: *Uber den Standort der Industrien, 1. Reine Theorie des Standorts* (Tübingen). English translation: Alfred Weber's theory of the location of industries, C. J. Friedrichs (Chicago 1929).
Weber, A. 1915: Zum Wesen und Zur Methode der Kultursoziologie. *Archiv für Sozialwissenschaft und Socialpolitik* 39.
Weber, A. 1920: Prinzipielles zur kultursoziologie (Gesellschaftsprozess, Ziviliza-tionsprozess und Kulturbewegung). *Archiv für Sozialwissenschaft und Sozial-politik* 47.
Weber, A. 1926: Kultursoziologie Versuche: das alte Aegypten und Babylonien. *Archiv für Sozialwissenschaft und Sozialpolitik* 55.
Weber, A. 1946: *Abschied von der bisherigen Geschichte: Uberwindung des Nihilismus?* (Bem). English translation: Farewell to European history or the conquest of Nihilism, R. F. C. Hull (London 1947). (This was a volume in the International Library of Sociology and Social Reconstruction, edited by Karl Mannheim.)
Weber, A. and Mitscherlich, A. 1946: *Freier Sozialismus* (Heidelberg).
Weber, Marianne 1975: *Max Weber: a biography* (New York).
Weber, Max 1975: *Roscher and Knies: the logical problems of historical economics* (New York).

Geography and Social Science: the Role of Patrick Geddes

B. T. Robson

The economic depression of the late nineteenth century is customarily dated from 1873; the heights of the current recession and stagflation date from the oil crisis of 1973. There is a strong temptation to draw parallels between the intellectual climates of the two periods so neatly encapsulated by this hundred year span. In the closing years of the nineteenth century, the erstwhile dominance of positivist orthodoxy in the prevailing political economy began to be challenged by the more consciously political flavour of the social sciences which developed after the turn of the century. Today, a similar challenge confronts neoclassical positivism as the social sciences have begun to develop phenomenological and structural approaches to tackle questions which positivist thought has either not asked or has demonstrably been unable to answer. Crude parallels of this sort, seeing in economic contraction the seeds of similar academic response, tread on slippery ground. The *fin de siècle* of one age cannot be that of another. Especially in the years since World War II Britain has undergone a fundamental change which has ushered in greater material affluence, greater leisure, falling profits and a growing disinterest in work for its own sake, a form of satisficing behaviour. The view that just as it was the first industrial nation, so Britain has begun to show the way towards the post-industrial age, is not without substance (Nossiter, 1978). If we change the terms and talk not of industrial and post-industrial, but of paleotechnic, neotechnic, geotechnic and eugenic, the terms introduced by Patrick Geddes no later than 1904, the argument that I want to explore in this essay becomes evident: that, in foreseeing some elements of the society that Britain was to become and in their analysis of the society with which they were then faced, some of the writers of the pivotal decades before World War I have methodological and practical relevance at the present day.

The political economy of the nineteenth century failed to provide a compass to guide social concern. What was needed was a more robust analytical approach to understand society, to recognize the conflicts inherent within it and to point to the way towards a better and a fairer alternative. In these two critical decades Geddes played a central, if eccentric, part and I want to consider the debate about social theory and the evolution of the social sciences, using, as a focus, his failure to unite the social sciences and the unrealized potential of some of his views.

As 'biologist, town planner, re-educator, peace warrior', in the words of his most recent biographer (Boardman, 1978), Geddes foresaw the buds of a new era which he wished to usher in. His task was to educate men to recognize and to develop their potential: 'as life transcends books, we may see, and, yet more, foresee, the growth of civic consciousness and conscience, the awakening of citizenship towards civic renascence' (Geddes, 1904b, 118). Unlike his contemporary, H. G. Wells, who also studied biology under T. H. Huxley, Geddes' ideas have been less scorned than ignored. Yet his nascent anarchism with its stress on the community and personal action, his recognition of the importance of the social structure rather than an atomistic empiricism in understanding society, his proselytizing for a united social science, his emphasis on evolutionary dynamics; all have relevance today and, despite their widespread neglect in the intervening years, many of his ideas now play a central part both in geography and the wider social sciences. That they do is no testament to his success as re-educator since his message was rarely understood, rarely acted on, and usually misinterpreted. He himself became a kind of academic Lloyd George, wandering in the wilderness. Rather, he paid the price of pioneering a multidisciplinary approach at a stage when the disciplines themselves were unformed. The stage is now more confidently set for the evolutionary and synthetic drama that Geddes foresaw as a middle way between the positive political economy of the capitalism of his day and the revolutionary tenets of Marxism.

If, at the end of the nineteenth century, political economy failed to provide answers to social questions, it was the new social science that was looked to as the alternative. If anywhere, here might be found the answers to the questions of why society continued to produce one-third of its people in poverty and how such a state of affairs might be improved. The nature of change and the direction of change were critical elements of these questions and the evolutionary ideas of biology together with an analytical approach to society might have produced just that combination which could have provided answers to the problem. In the event, rather

than a unified social science, a congeries of disciplines were to develop
and they were not to be characterized by analysis and theory.

I want to consider how it was that geography and planning, in particular, were to develop as they did, turning their backs on theory and
postponing, until our own day, the development and incorporation of
social theories. I shall first look at the background of political economy
and the fission of the newly developing social sciences and then at some
of the unrealized contributions inherent within the debate about society.

SOCIAL CONCERN

Political economy and ameliorism

It would be wrong to see the political economy of the middle years of the
nineteenth century, out of which the social sciences were to emerge,
as being monolithic and unitary. Nevertheless, the prevailing positivism
of its approach, whether in the hands of its statisticians or those
ameliorists concerned with social reform, led to a remarkable agreement
as to what was needed by way of research. The underlying belief in a
rational solution to society's ills, in a view of consensus and social
harmony and in an invisible hand which would guide society towards
greater prosperity and social welfare, meant that what was needed was
information, pure and simple, unguided and unpolluted by theory. Disputes would disappear if men were but presented with the 'facts' and an
apolitical presentation of those facts would necessarily lead to a better and
more efficient policy, a legislative programme for redressing social ills. In
such a framework, as Abrams (1968) argues, the chance for a genuine
analytical sociology to develop were frustrated at every turn. The
emphasis on the individual rather than on social structure diverted attention away from a sociological explanation and most social investigators
saw their work as feeding directly into the formulation of legislation
rather than of academic theory. Intelligence was reduced to numerical
tabulation. Indeed many of the investigators were themselves legislators
or politicians or had aspirations to be such. The work of Charles Booth has
to be seen in precisely this context as a massive compilation of facts –
unguided by theory and not leading to theory – out of which policy was
misguidedly thought to stem naturally and inevitably. The statistical
societies, led by Manchester and soon followed by London, and the social
surveys of the charity workers followed just such a philosophical tack
within the framework of the progress and economic expansion of the
middle years of the century. Poverty, in such a view, was not to become a
structural condition, not an explanation, but a mere context within which
depravity and drunkenness in the individual were heightened evils and
barriers to his moral development.

The fission of the social sciences

The growth of a more truly analytical social theory developed from the evident inability of the facts to speak for themselves, out of the growing realization that policy could not unambiguously derive from neutral surveys. It is not without significance that it was the growing economic crisis of the 1870s that provided the context for the re-emergence of a more politically charged theory. The strands which emerged have been characterized by Halliday (1968) as a trilogy comprising ethical sociology with a focus on social work, racial sociology with a basis in eugenics, and civic sociology with the evolutionary and synthetic views espoused by Geddes and his associates. Just as, earlier, the different emphases within political economy revolved around the separate institutions of the Statistical Society and the National Association for the Promotion of Social Science, so the emergence of these three schools was, at least at the outset, embodied in the institution of the Sociological Society, established in 1903 by Geddes' friend Victor Branford as a forum for Geddes' views and as a publisher of a journal of civics. The three volumes of *Sociological Papers* which were the published fruit of this body, and which gave birth to the *Sociological Review*, embody all three strands: the sociologists by L. T. Hobhouse and E. Westermarck, the eugenists by Francis Galton, the school of civics by Geddes and Branford, the latter being secretary of the Society. The only geographer, as such, who was elected to the council of the new body was H. J. Mackinder whose involvement appears to have been somewhat peripheral, given his preoccupation both with establishing geography as a discipline in its own right and with guiding the London School of Economics of which he had become director in 1903.

The background out of which the Sociological Society might have been able to weld a unitary analytical social science, distinct from economics with its monetary basis of value and from anthropology, was essentially the evolutionary theories spawned from Darwin and developed through the writings of Comte and Spencer. The differences between the three schools, however, proved too strong for any such fusion to occur. Instead of an interdisciplinary social science which might have helped to tackle some of the problems sidestepped by political economy, the separate disciplines of sociology, geography and demography were to emerge and to develop their associated institutions, teaching syllabuses, viewpoints and emphases along with their inevitable trappings of professionalism. Simultaneously, planning was to crystallize as a further separate discipline alongside the first halting steps towards effective town planning legislation.

Such fission was principally the result of the different interpretations of the still hazy science of biological evolution. The eugenists' emphasis on the pre-eminence of heredity and their consequent concern about differential fertility and what they saw as the diminution of the nation's health stock, led them to an advocacy of differential birth control and a crudely Darwinian eugenics. With the establishment of the Eugenics Laboratory and the Eugenics Education School, the views of Galton and Karl Pearson pulled further and further apart from those of other social scientists. The eugenics debate became increasingly conducted through more and more sophisticated statistical techniques and, to this extent, continued the earlier tradition of political economy, with the *Journal of the Royal Statistical Society* acting as a forum, to which now were added the *Eugenics Review* and *Biometrica*.[1] In contrast to this, the holistic view of Geddes' civics school was that the abstraction of either environment or heredity from the interacting whole of environment/function/organism was illegitimate. Biological mechanisms such as natural selection and the survival of the fittest could not be applied by analogy to derive social 'laws' since this was to divorce man from mind. The distinction between eugenics and civics can thus readily be established. That between civics and ethical sociology is somewhat more difficult. Hobhouse was concerned to develop a theoretical sociology which built on Spencer's evolutionary concepts without recourse to the latter's political views of the undesirability of intervening in the pace of evolution. Hobhouse's solution was to stress the evolution of the human mind such that, with the coming of self-consciousness and increasing control over his environment, man was better able to participate in his own destiny, to control it and therefore to erode the iron rule of evolution. Many of his views on the evolution of mind were similar to Geddes', but the latter was impatient both with the fact that Hobhouse's aim was a theoretical one and that his stress was on the sufficiency of purely rational thought. In practice, the ethical sociologists were to lead to the establishment of social work and social administration which Geddes would have considered insufficiently broad ranging. His incorporation of environmental influences, his view of the irrational and individual interpretation of the dynamics of evolution, and his emphasis on the need for personal self-awareness, self-improvement and involvement could never adequately be subsumed by the training of social workers.

GEDDES' VIEWS

Geddes' own views are difficult to summarize, stretching as they do over so broad a field, but the principal doctrines which sustained them and the philosophical framework in which he developed them are clearly organic and evolutionary, unitary and vitalist. His two best known writings – *Cities in evolution* and *Civics as applied sociology* – need to be read and interpreted through this philosophical position. His antipathy to organized education and the formal trappings of university requirements and degrees, his personally adopted motto of *Vivendo discimus* and its implied stress on personal involvement and experience, his constant urging of the unity of matter and of knowledge, his stress on dynamics and the flow of change with his consequent study of cities through sequences of things past, present and to come, can all partly be traced to his early biological education:[2] but they derive from roots deeper and more complex than formal training. Certainly the nature of Edinburgh itself played no small part in the development of his interests and his approach, just as in so different a way Chicago was to underpin and generate so much of the specific focus of R. E. Park. The fact that it was a town of only modest size whose lineaments could be absorbed and comprehended in one view from Arthur's Seat, or later from his camera obscura in Outlook Tower, and that its poverty and squalor, though real enough in the Scottish tenements of the Old Town, were nevertheless on a small and individually comprehensible scale, clearly lent the town to a humane rather than a mechanistic, organic view in a way that the size and disorganization of a London or a Chicago might not (Kitchen, 1975, 42). Equally, the wealth of historical associations embodied within Edinburgh was important to his emphasis on dynamics. The golden age of encyclopaedists formed an essential prop to his view of the pageant of history and the concatenation of circumstances through which such greatness might again arise.

The immediate stimuli from which Geddes fashioned his views are Ruskin, Comte and Le Play. The latter in particular provided a method, a means of organizing some of his ideas. Le Play's view that the environment and mode of livelihood of an area is reflected in common social characteristics and particularly in the *mores* and the nature of the institution of the family, provided the basis of Geddes' trilogy of place/work/folk. Equally the survey was to become his principal method. Le Play had concentrated on the family, arguing that the onset of industrialization had destabilized the patriarchal family and that in the so-called 'stem' family could be found a form which should be reconstituted to meet the challenge of the industrial age. Geddes widened the focus from *famille* to 'folk' and set his sights more broadly than the detailed family budgets which formed the kernel of Le Play's social surveys. The social survey and interactions between environment, function and organism formed the essence of Geddes' method. But his civic surveys were not to be the factual 'snapshot' ones of Booth, which still, as Geddes saw it, left London 'a foggy labyrinth' (Geddes, 1904b, 104); rather, their object was to set the citizen within the stream of influences derived from the history and the nature of the town in which he lived.

This emphasis on time and evolution came from Comte, from whose holistic and organic sociology and Law of Three Stages Geddes derived his evolutionary approach and his use of stages in time. His view of the city as 'drama in time' is a complex multidimensional one. It combines a

regional geographical element – the 'river section', which itself embodies
both the dimension of time through the evolutionary sequence of technological development of an increasingly complex society as well as the
influence of spatial dependence – together with the time dimension itself
– the pageant of history embodied as a palimpsest within the accretions of
a given city. Tyrwhitt (1949, xi) sees this as a dual vision from a composite
mind. The blending of geography and history was certainly not understood when the paper itself was given to the Sociological Society, as the
comments of Booth and others make clear (Geddes, 1904b, 119–44).
Geddes' sequence of stages is correspondingly complex. They cannot be
seen as some inevitable teleological progression, fuelled by technology,
in the way that Mumford was later to translate the ideas (Mumford, 1934),
since Geddes distinguished between a mechanical progression of the
hunter/pastoralist/agriculturalist variety and his own view of the complex struggle between competing 'social formations' in which any one
stage is the resultant of all the preceding and new types of formation
(Geddes, 1905, 61) – the struggle, for example, between the social formations represented by 'renascent Florence' and 'decadent Versailles'
(Geddes, 1904b, 114). In this essentially structuralist view, the inspiration, inarticulate and chaotic as it is in his writing, is clearly organic and
vitalist, a view of man as being infinitely able to renew himself and
transcend himself. Here is something of the kernel of what was to become
a more clearly, if never very convincingly, articulated doctrine in the
writing of the French philosopher Henri Bergson (1911) and of Geddes' ex
fellow student Lloyd Morgan (1923).[3] The importance of this vitalist
strand forms a crucial part of Geddes' thought (Mellor, 1977). The *élan
vital*, the subjugation of intellect in favour of instinct and intuition — an
emphasis which was to lead Russell (1946, 821) to say that instinct was
'seen at its best in ants, bees and Bergson' – and the view of change as a
continuous unbroken stream of 'becoming', are all an integral part of
Geddes' dynamic view of citizenship and civics. It led to the almost
mystical aspects which were to become increasingly prominent in his life
after 1914 in India and France. It led, too, to much of the difficulty he had
in communicating his ideas to audiences who would sit, 'a little dazed
perhaps, listening to endless variants of the Place–Work–Folk theme'
(Fleure, 1953, 9).

One of the keys to understanding the contrapuntal way in which his
mind worked and the elusive richness of his writing is through his
'thinking machines' which he used to expound his ideas to the obvious
bewilderment of his listeners.[4] His squared pieces of paper, first developed during a period of temporary blindness in 'Mexico, suggest interconnections which were never spelled out analytically in verbal terms,
reflecting yet further the intuitive anti-intellectual appeal of Bergson. In
some instances of these 'thinking machines', for example, he traces the

correspondence of geography, economics and anthropology as the disciplinary equivalents of place, work and folk or of environment, function and organism. The cross tabulations of each element of the trilogy produce such combined elements of his matrix as work-FOLK, the social structure based on occupational classification, or folk-PLACE, the home and locality of a region's inhabitants (Geddes, 1924). In others, he interprets his large-scale view of the relationship between the original formula of Le Play and the civic goal to which he aspired through a sequence of inverted and transposed matrices; leading from the everyday world of the town, through the thought world of the school, and the contemplative work of the cloister, to the synthesis of the city (Geddes, 1905). In these diagrams – his personal version of a matrix algebra – he moves at random up, down, across and diagonally in hinting at the connections. 'The application of this formula,' he disarmingly notes, 'will not be found to present any insuperable difficulty' (91).

The difficulties, of course, were innumerable. His evolutionary theory was a compound of visionary wishful thinking and organic reification of the city. This is not to say that it did not contain perceptive insights, but they came as much from his intuitive as from any analytical approach. And the practical tool which was to demonstrate its feasibility, the social survey, was singularly ill-specified: he never developed a method by which such surveys might be conducted; and he was always considerably clearer in suggesting the historical, land use and occupational data that should be collected than how one should collect information of a social kind. Such town surveys could only too readily become, in other hands, mere mechanical compilations of old buildings and local history. But this is to ignore his real aim which was educational rather than analytical. His surveys were to open men's eyes to what *he* knew was so and to teach them what *he* knew was the solution to urban problems. Le Play's influence in Britain, carried by Geddes via Demolins, was thus to be of a more heuristic than analytical nature (Brooke, 1970; Mogey, 1956). Improvement, progress, the solution to current ills and the way towards the eugenic age lay in the improvement of man; in the combination of reflection, thought and culture; of the synthesis of ideal, idea and imagery; of the final synthesis of the culture palace of the cathedral. And the survey was the lever to the new age. The dawn of Geddes' new world lay in the awakening of the citizen; his own role was to teach them to see for themselves, to learn for themselves and to act on their own initiative. *Cities in evolution*, said Mumford (1956, 105), 'was not a contribution to scholarship: its essential subject was the education of the citizen towards his understanding of urban processes and his active assistance in urban development.' It is in this light that one can understand Geddes' claim that his civics was 'concrete', and that one can understand his enthusiasm for arranging pageants, encouraging local artistic tradition in the spirit of

William Morris (and his delight at the widespread acknowledgement in Europe of the work of the Edinburgh architect Mackintosh) and, above all, for the public exhibition. It was this latter that gave him greatest scope to put into practice his contrapuntal approach to amassing and organizing knowledge. He used the opportunity to display plans, maps and photographs in precisely the same way as he 'read' his thinking machines, as is evident from his account of the logic of his Ghent Exhibition in 1913 (Geddes, 1915, 270–89). As with the current fad for poster displays in academic meetings, such exhibitions combine some analysis with a great deal of appeal to intuition. Only with Geddes himself as guide to and interpretor of his exhibitions or his Outlook Tower must the static mute material have begun to take life. Abercrombie (1933, 128–9) suggests as much in his description of Geddes' enormously successful display at the Town Planning Exhibition of 1910:

> It was a torture chamber to those simple souls that had been ravished by the glorious perspectives or heartened by the healthy villages shown in the other and ampler galleries. Within this den sat Geddes, a most unsettling person, talking, talking, talking . . . about anything and everything. The visitors could criticize his show – the merest hotch-potch – picture postcards – newspaper cuttings – crude old woodcuts – strange diagrams – archaeological reconstructions; these things, they said, were unworthy of the Royal Academy – many of them not even framed – shocking want of respect; but if they chanced within the range of Geddes' talk, henceforth nothing could medicine them to that sweet sleep which yesterday they owned. There was something more in town planning than met the eye!

THE UNREALIZED CONTRIBUTION

The obscurantism and the mystic anti-intellectualism of Geddes, then, might be said to have contributed to the failure to develop a united social science out of the politically heightened promise of the late nineteenth century. Yet not only in the strength of his own vision and the charisma of his person, but in much of his actual writing, there are insights and approaches which we have only begun to re-learn two generations later. Like that other vitalist, Bernard Shaw, his prescience was at times remarkable. His one successful coinage of a term – 'conurbation'[5] – is perhaps that for which he is superficially most widely known. Behind the term lay an acute analysis of the growth of city regions. Like Mackinder's view of the influence of metropolitan London or H. G. Wells' vision of the spread of metropolitan growth, and long before Jean Gottmann, Geddes foresaw in North America the 'vast city-line along the Atlantic for five hundred miles, and stretching back at many points, with a total of, it may be, as many millions of population' (Geddes, 1915, 48–9). His optimistic forecasts of the beneficial influence of the neotechnic age may have been doomed not to be realized since most urban development simply built on the existing base of the old paleotechnic order, much no doubt to the dismay of the fairy godmother of his delightful Cinderella parable 'waving her fairy electric wand as the herald of the new era' (Geddes, 1915, 129). But, besides such prophecies, whether prescient or foredoomed, I want to discuss five particular contributions that Geddes made. Each was un-realized in his day, but has now become part of our current wisdom.

Community: small scale and anarchism

First is his emphasis on the community, with all that this entails in terms of a preference for the small scale, for participation and for the anti-governmental views which are implicit and occasionally explicit in his writings. Two strands of this advocacy can be disentangled. First is his stress on small-scale rehabilitation, of accommodating to what is and adapting and improving where possible rather than having resort to large-scale, perhaps utopian, plans on the drawing board. He was con-stantly sceptical of the T-square solutions of the architect, critical of Hausmann's style of urban planning both for its overly formal, overly grand designs which smacked of militarism and manipulations: 'The great city is not that which shows the palace of government at the origin and climax of every radiating avenue: the true city . . . is that of a burgher people, governing themselves from their own town halls and yet express-ing also the spiritual ideals which govern their lives' (Geddes, 1915, 254). He was in essence more sympathetic to the town, less anti-urban, than were the architects and planners who were his contemporaries. It is here that he needs to be distinguished from Ebenezer Howard and the advo-cates of the garden city movement since, while he agreed with their emphasis on low density and he himself advocated the decentralization of industry, he mistrusted the impulse to redesign towns *de novo*: 'We have to live in towns: and on the whole, with respect to the Garden Cities and Garden Suburbs, we have to make the best we can of the existing towns' (Geddes, 1915, 220). Such a view is a product both of his evolu-tionary view of the future growing directly as the resultant of the con-flicting forces of the present and of his mistrust of the large scale. In his own practical work and planning advocacy the same impulse is evident. In Edinburgh, he was concerned to establish small public gardens and calculated the amounts of land within the city which were available to meet such purposes. In India, he poured scorn on those who swept away the small village water tanks, and urged instead the use of a film of oil which would preserve the tanks yet inhibit mosquitos. Such 'conservative surgery' as he called it is much in tune with our own contemporary pressure for rehabilitation and preservation in place of large-scale renewal.

The second related strand is his advocacy of self-help and self-government. His mistrust of formal government (which is not inconsistent with his advocacy of regional planning for his 'new heptarchy' of conurbations), his approval of the tenants' co-partnership schemes being set in train by Henry Vivian (Geddes, 1915, 138–9) and the whole drift of his 'concrete' sociology as an exhortation to citizens to involve themselves in their own destiny, all point in the same direction. His own activity in his Edinburgh Social Union – through which small-scale improvements were effected to the squalid environment of the Old Town, buildings were purchased for repair and University Hall was established as a residence in which students themselves participated in its self-government – testifies to his personal commitment. By 1896 when the Town and Gown Association took over most of these responsibilities from him the list of assets included 4 University Hall houses, 85 re-conditioned or newly-built houses and a number of shops and building sites (Mairet, 1957, 76). His dislike of burgeoning government and its creeping bureaucracy is part of this same self-help thrust. Rather than rely on tax-financed central government intervention, he urged his fellow citizens to 'use the money themselves in the first place. Why not keep your money, your artists and your scientists, your orators and your planners – and do up your city for yourself?' (quoted in Defries, 1927, 221). His aim, that his ideas would lead to 'experimentally opening our eyes towards that substantial Resorption of Government, which is the natural and approaching reaction from the present multiplication of officialdom' (Geddes, 1915, 101), sets him clearly as sympathetic to, if not actively part of, an idealist anarchism which is close in spirit to the community action and participation which has more recently emerged in the face of corporate capitalism and central planning. It is no accident that the anarchists Peter Kropotkin and Elie and Elisée Reclus (see chapters 8 and 9 of this book), attended some of Geddes' Edinburgh summer schools in the 1890s.

Social structure

Second, one can point to the structuralism incipient in Geddes' thought. What marked his originality as a potential social commentator, and what equally marked some of his contemporary sociologists, was that he eschewed the atomistic and empirical work of the earlier nineteenth-century political economists. Le Play, of course, was a valuable guide in this respect since he *was* a sociologist who combined a drive to ameliorate social ills with an analytical approach to his view of society. Ruskin, too, in his critiques of bourgeois society saw the social order in analytical terms rather than purely monetary or moralistic. Geddes' first two published essays in social comment clearly suggest the force of his analytical approach. In his 1881 paper, received with some bewilderment when it was delivered to the Edinburgh Royal Society, he developed a system for cultural studies which was based, not on the material categories of the economists, but on functional categories based on social organization; and in his 1884 paper he further developed this theme by advocating the substitution of economic man by biological man in interaction with his environment. In both papers he went beyond the much simpler descriptive categories which were to be used, for example, by Charles Booth. Instead, he worked towards both a sociological application of the modern concept of energy, in his use of man-power or man-days, and a definition and application of social indicators 80 years before they were eventually to be used formally in social science (Abrams, 1968, 114). His view of social structure is a sophisticated one, with more than passing resemblence to a structuralist approach.

Change could be seen as the resultant of warring social formations which 'combine with, transform, subjugate, ruin and replace each other' so that he could 're-interpret the vicissitudes of history in more general terms, those of the differentiation, progress or degeneracy of each occupational and social type' (Geddes, 1905, 60). He traced the rationale of social institutions to the underlying structure of such social formations, as in his outline of the school curriculum with its component parts being the product of the material or political demands of different formations: the three Rs as a product of the need for cheap clerical labour in the age of finance, grammar and 'the precise fidelity of absurd spelling' as an imitation of the proof readers of the renaissance, the essay as the abridged form of medieval disputation (Geddes, 1904b, 112). And his analysis of the paleotechnic age, in *Cities in evolution*, clearly demonstrates his view of the structural imperative of the society whose demise he wished to hasten. Three quotations (Geddes, 1915) encapsulate the essence of this view:

> As paleotects we make it our prime endeavour to dig up coals, to run machinery, to produce cheap cotton, to clothe cheap people, to get up more coals, to run more machinery, and so on; and all this towards 'extending markets'. The whole has been essentially organized upon a basis of 'primary poverty' and 'secondary poverty'. . . . (74).

> The paleotechnic order should, then, be faced and shown at its very worst, as dissipating resources and energies, as depressing life, under the rule of machine and mammon, and as working out accordingly its specific results, in unemployment and misemployment,

in disease and folly, in vice and apathy, in indolence and crime. All
of these are not separately to be treated . . . but are logically connected . . . they are worked out, in sequent moves, upon the chess-
board of life (86).

. . . within these narrow beats, these essentially parallel streets, in
which 'upper' and 'lower' class matter so much less than either
thinks, the mind of the capitalist and his political economist, of the
labourer and his economist, of all their woman-folk, are alike half-
blanched, half-blackened – grey lives all (128).

I am not suggesting that Geddes ever developed a formal theory of the
bases of the structural order in the society of his day. What is clear,
however, is that here were the elements of a powerful social theory; one
which Geddes himself was to turn away from in his progressive shift
towards planning and which was not to inform geographical writing for
another two generations.

Holism

Third are some of the offshoots of the holism of his writing. I am less
concerned with the 'emergent properties' that are an integral part of
holism, and which seem difficult to define analytically by their very
nature, as is apparent in the difficulties faced by the regionalism of La
Blache, with whose work Geddes was highly impressed. What does
come from his holistic breadth of interests, however, is, first, his multidisciplinary approach to cities and, second, his concern for what now we
would call environmental management. The first clearly anticipates our
current concern with multidisciplinary approaches to urban studies and
Geddes' work in this regard was certainly out of tune with the development of fragmented disciplines. 'Instead of perceiving that Geddes'
capacity for co-ordinated and interrelated thinking was a far more exacting discipline than any single specialism demanded, Geddes' conventional contemporaries saved their own pride by treating this special
capacity as inferior to their own smaller gifts' (Mumford, 1956, 102). In the
second, his concern about regional networks of water supply, the creation of parks, the impact of the paleotechnic era on resources and the
wiser conservation heralded by the new ages, are all a clear foreshadowing of topics in resource use and management and, too, in environmental
perception which now figures so prominently in geographical work.

Dynamics and evolution

A fourth element which now finds explicit echoes in current geographical
work is the key role of dynamics in his writing. Movement he saw as the

essence of life and of the societies which he studied. A recurring criticism which he levelled was that 'we have come to think of this present type of town as in principle final, instead of in itself in change and flux' (Geddes, 1915, 362). The anabolic and katabolic functions, about which he wrote in his work on the biology of male and female, permeate his view of the struggle of competing social formations and his vitalist philosophy led him inevitably to the view that decay was a necessary prelude to rebirth and possible improvement. His constant emphasis in his advocacy of the city survey was that one should study the present and the historic pageant of the city merely as a prelude to the projection of the city of the future. Static cross-sections form no part of such a view. Again, after the preoccupation with the static equilibrium models of geography's recent past, this view – that 'the city is more than a place in space, it is a drama in time' – strikes a very contemporary note.

The urban focus

The fifth, and final, point which can be urged is his very focus on the city itself. Mumford was to comment, 'No member of the present generations, with normal opportunities for study and reading, can fancy how exciting any book on cities was before 1920 – to say nothing of one like Geddes', which related the transformation of cities to the social, economic, and cultural situation of our time' (Mumford, 1956, 105).[6] While Mumford himself was to carry this urban and vitalist creed to America as Geddes' most faithful acolyte, in Britain urban texts indeed continued to be few in number. Those which appeared in geography were either unashamedly historical (Cornish, 1923) or statistical and descriptive (Fawcett, 1922). Later texts like Dickinson's (1947) were more influenced by American and German ideas than by Geddes and his French influences, since Dickinson drew his inspiration more from central-place concepts and the human ecology of Chicago than from any structural/functional sociology. What Geddes might have contributed was a solidly-based urban sociology; instead, ironically, the emphasis on his style of city surveys led to a non-sociological form of planning surveys and, in Glass's view (Glass, 1955), their diffuseness and eclecticism discredited urban sociology in Britain. There was no practical and disciplined interpretation of the chaotic urban vision that Geddes offered. Yet, again, the potential is readily apparent, not only in his writing, but in the syllabuses of his lecture courses and summer schools. Had university teaching in geography followed more closely both the letter and the spirit of his course at Oxford in the summer of 1910, for example (*Geographical Teacher*, 1910), a stronger seedbed for something akin to modern urban geography would undoubtedly have been laid. Its blend of description, of sociological interpretation, of history and of social roles, embodies in embryo much of the now-grown tissue of urban geography.

THE DILUTED LEGACY

Given these potential contributions, the direct legacy of Geddes comes as an anticlimax. An analytical urban sociology did not develop; social science became fragmented into separate disciplines; dynamic and evolutionary planning did not emerge. The social concern of the late nineteenth century was not to find the theoretical structure which could inform social policy. Instead the gradual emergence of the welfare state was to be founded on parliamentary blue books and commissions which continued the spirit of political economy, of gathering facts within a theoretical vacuum. Geddes' direct legacy to the inter-war period reflects this failure. It was the bones and not the spirit of his early thinking which were transmitted. His organic evolutionary theory, his analysis of social structure, his plea for community action and development, his mysticism, were largely lost. What were passed on, to form a thin gruel, were his survey approach and his interest in regions. These elements were transmitted directly to planning and to geography and, in both cases, their effects proved deleterious.

Planning

Geddes played a central role in the early establishment of planning. Not only was he acquainted with Raymond Unwin and other leading advocates of planning who helped to promote the passing of the first town planning act in 1909; he was involved in the founding of the Town Planning Institute and was created its honorary librarian, without, as Cherry (1974, 58) notes, there being any evidence that the Institute had a library. Geddes' injunction, 'survey, analysis, plan', became an article of faith to the planning profession and his style of surveying contributed much to the method of the profession, especially in the work of Abercrombie, whose reports in the 1920s and 1930s preceded his two influential London reports. Geddes' plan for the Carnegie Trust in Dunfermline (Geddes, 1904a) helped to establish his reputation as a planner and he assiduously cultivated his national and international contacts which helped to provide support from his planning exhibitions. The Dunfermline report, with its characteristic illustrations in 'before and after' photographs and its advocacy of a landscaped park, scattered with palaces of culture, which would act both as lungs for the town and as pointer for the spiritual uplift of its citizens – the realization of his 'cathedral' – provided not only a model for his own later reports of Indian cities, but also influenced the style of surveys and plans from the newly evolving planning profession. Abercrombie claimed to have 'read it over and over again, and used it as a basis for other reports which I have engaged on myself' (quoted in Boardman, 1978, 447).

Even here, however, there are subtle differences in the intent of

Geddes' early work and their translation into the profession of planning. His recommendations for Dunfermline were that the Trustees invest Carnegie's capital and use the interest to effect a continuous series of small improvements – a style of continuous dynamic planning which has only become central in British planning since 1968. His welcome for Burn's planning Bill was based on its conservatism (Geddes, 1915, 207–8) in contrast to many of the contemporary planners who wished to see less hesitant powers being given to local authorities. British planning was to become strictly physical and essentially static. The spirit of Geddes' advocacy, in contrast, was social and dynamic. Expertise, not social vision, inevitably came to characterize the profession as technique and skills became passports for entry. 'From now on the course of town planning was to be very considerably affected by professionalism and the liberal, humanitarian, and social reform element was partly squeezed into a professional frame' (Cherry, 1974, 61). Geddes' campaign for civics was to become 'a beautiful backwater as far as the future of both sociology and town planning were concerned' (Mellor, 1973, 315).

Geography

As in planning, so in geography Geddes' influence was considerable, and equally it was the bare bones, not the spirit, which predominated. Given his wide contacts and his own emphasis on geography, it is not surprising that Geddes knew, and influenced, a number of geographers. Mackinder one has to see as an equal, a man with a reputation already established by the turn of the century when Geddes began to make a wider impact. That Mackinder was to play so peripheral a part in the evolving social sciences and, following the tradition of the political economists, was soon to move into politics, was one of the lost opportunities of the period. It was through A. J. Herbertson and H. J. Fleure, however, that the Geddes legacy was more particularly felt within geography. Fleure was one of the team which Geddes took to Dublin in 1914 and with whom he maintained an active contact. What Fleure took from Geddes was his synthetic approach which fitted well with the then current regionalism of French geography. Fleure's interest in the inseparability of man and environment and his concern for international co-operation have much of the spirit of Geddes, and his own initial training in zoology provided for him a background not unlike that of Geddes. His interests however were, and increasingly became, in archaeology, physical anthropology and pre-history. His evolutionary quest was applied more to an understanding of the past for itself rather than as a prelude to a forecast of the future and the amelioration of present ills; a view of time very different from Geddes' notion of the continuous flow of becoming. Herbertson was assistant to Geddes at Dundee University and one of the nucleus of close helpers who sustained the Edinburgh summer schools. He was the second direct

strand in the connection with academic geography.[7] More strongly
determinist than Fleure, his development of the regional concept is his
strongest academic contribution, but in this his almost wholly physical
definition of environment did less than justice to the holism of Geddes.
Both Fleure and Herbertson were deeply involved in furthering
geography as a teaching vehicle. Fleure was to succeed Herbertson as
Secretary of the Geographical Association in 1917 after the latter's death
in 1915 and in this position they were both centrally placed to advance the
subject's role in schools – in Fleure's case, against the added urgency of
the threat to remove geography as a main subject within the Higher
School Certificate (Garnett, 1969). To this end, they both continued
Geddes' emphasis on summer schools; Herbertson in Oxford from 1906
(Gilbert, 1972, 192) and Fleure in Aberystwyth. This involvement in
teaching and in the work of the school-orientated Association obviously
reduced any research impact that Herbertson specifically might have
made and gave to the subject generally a non-research emphasis which
equally made unlikely the crystallization or the development of Geddes'
ideas. Add to this the fact of Herbertson's physical interpretation of
regionalism, and the dilution of the Geddes legacy is readily understood.
Dickinson's (1969, 204–5) view would be widely subscribed to: '. . . these
trends, initiated in the work of Patrick Geddes . . . may have had
deterrent effects on the development of geography (as for example in the
continuance of "regional geography" and the continual use of the "geographical background" as the physical elements only of the area in
question).'

If the framework within which geography worked was the region, the
chief teaching vehicle and research tool was the survey. Again the in-
heritance was from Geddes, and it was formally established under the
auspices of the Le Play Society, which was spawned as an offshoot of the
Institute of Sociology and was concerned with conducting a 'vigorous
campaign for the prosecution of field studies of the Le Play-Geddes
model, and for the spreading of "regional survey" amongst the schools
and training colleges' (Beaver, 1962, 234). In its work, Dickinson, K. C.
Edwards, Estyn Evans, Fawcett, Fleure, Ormsby, Stamp and many other
geographers both learned and taught, and helped give academic geography the strong field work and survey basis which so characterized it
even in the 1960s. While the influence of Geddes in this cannot be
doubted, it was yet again an influence which can be argued to have had
baneful effects, partly because, set within the regionalism of the day, the
ill-specified nature of Geddes' survey technique all too easily led to an
emphasis on measurable, and particularly physical, aspects of an area
and to surveys that were merely wide-ranging descriptive accounts.
Partly, too, it was again that the spirt and aim of Geddes' mission was
lost. His social crusade, his goal of social awakening and betterment,

evaporated in the general descriptive accounts which came from this geographical translation of his method. A contemporary verdict on the surveys of the inter-war period usefully summarizes the limitations both of the survey itself and the regionalism to which it contributed:

> Regional surveys, it is probably true to say, do not claim yet to have progressed beyond sketches of the geographical and historical 'substructure'; and the published work of this character contains little attempt, so far as highly developed countries are concerned, to relate these data with those on contemporary society. While therefore this method has had much value in the hands of Le Play and his followers for the study of less developed societies, its adequacy for the treatment of a highly developed industrial area in England remains doubtful (Wells, 1935, 48).

Only with a more rigorous philosophy or a more narrowly defined and substantive goal than the region could Geddes' groping towards an urban 'sociography' (Geddes, 1905, 66) have been realized.

CONCLUSION

How can this diluted legacy, this baneful effect, be explained? How can the promise of the social debate at the turn of the century have led to non-sociological, static planning, to descriptive regional geography and to the absence of a robust urban sociology? First it is clear that the separately evolving subjects were more concerned to develop their own identity and distinctiveness and that the uncertainties of the science of biology and evolution provided scope for diverging views of how evolutionary concepts could be applied to the study of society. The 'disciplinary drive' is clearest in the case of planning where professionalism and the desire for demonstrable technical expertise re-inforced the tendency for planning to become a matter of physical design and of technique. But in geography, too, the establishment and expansion of the subject dominated its evolving nature. The emphasis on schools and on teaching was helped by the fact that the survey and field work could as readily be turned to descriptive uses as to analytical. In neither subject was theory a prerequisite for the desired expansion so long as technique or method could be used as substitute.

Second, so far as Geddes himself is concerned as one of the strands in the social debate, his impact was inevitable blunted by the unrefined nature of his message, the anti-intellectualism of his reliance on thinking machines and his eventual resort to propaganda in place of analysis. His use of surveys demonstrates these weaknesses since, while ostensibly analytical, they largely demonstrated, not explored, what he already believed. They could not be used analytically unless they were to be developed with a substantive narrower focus. One example of what they might have become – a specialized substantive survey – is the survey of housing in Manchester by T. R. Marr, who had run Outlook Tower until 1901, had been submerged in the chaotic finances of Geddes' Edinburgh commitments and was later to be associated with the Collège des Ecossais at the end of Geddes' life. His survey (Marr, 1904), while primitive in terms of its sampling design, led to highly prescient recommendations in the housing field, including the greater involvement of local authorities in the building of houses and the ownership of land, the rating of un-developed land and the establishment of low-interest loans. Only by the translation of some of Geddes' ideas into substantive fields of this sort could the survey, the focus on towns and the ameliorative intent of his message have been translated into a more robust legacy. Had there been a disciple more 'willing to arrange the dried flowers of Geddes' thoughts into lifelike bouquets' (Mumford, 1966, 18); had his ideas been less diffuse, less messianic; had his civic survey been translated into a method which viewed social phenomena with the context of a structural view of society: then his message would have lost some of its crusading essence, but might have had a more beneficial and direct impact. Development of the ideas arising from the debate at the turn of the century has had to await the prior entrenchment of separate disciplines in the social sciences. Now that many of his early ideas have been independently rediscovered and incorporated in the more united social sciences and in social welfare, we are perhaps in a stronger position to appreciate the unrealized poten-tial of the 'wandering scholar' whose self-imposed brief was not merely to plan the city but to plan the world.

Notes

[1] The parallel with recent developments in quantification in geography is striking. In the face of the philosophical debate about positivism, many of the quantifiers have retreated into the stratosphere of higher mathematics; just as, in the earlier period of debate 'methodology apart, the statisticians stood still while the universe of social enquiry grew around them' (Abrams, 1968, 30).

[2] It is interesting to trace the very different application of biological ideas in American social science where Park was to develop his biological analogies in a crude Darwinian fashion of which Geddes would have thoroughly disap-proved. In their *Introduction to the science of society*, Park and Burgess note Geddes' *Cities in evolution*, but only as a mere example of a local social survey, they also note his work in biology with J. Arthur Thompson. W. I. Thomas, who in the period 1896–1919 was a member of the sociology department at Chicago and 'became its most creative and influential member during that period' (Faris, 1967, 13), dismissed Geddes' early paper to the Sociological Society by suggesting that, 'From the standpoint of its applicability to new countries like America, Professor Geddes' programme is inadequate because of

its failure to recognize that a city under these conditions is formed by rapid and contemporaneous movement of population, and not by the lapse of time' (Geddes, 1904b, 135). This may have been true, but the more fundamental reason for the lack of impact of Geddes' work on the human ecology of Chicago was that the respective views of the applicability of biological knowledge were radically divergent.

[3] This vitalist doctrine permeates the somewhat uninspired writings of Branford who published work both in collaboration with Geddes and on his own account in books and in frequent articles in *Sociological Review* of which he was for long editor. He probably compiled the bulk of the reports of the Cities Committee of the Sociological Society which was set up at Geddes' instigation. This Committee began to publish a series of reports, the first appearing under the title 'Towards the third alternative' in *Sociological Review* 11, 1919, 62–5. A typical example of the vitalist flavour is given in 'Body, mind and spirit', *Sociological Review* 14, 1922, 1–23.

[4] Booth, chairing the meeting, was to dismiss these diagrams as 'charming' (Geddes, 1905, 112). The use of the 'thinking machines' is perhaps most clearly suggested in an appendix to the revised edition of *Cities in evolution* (Tyrwhitt, 1949) and, particularly, in the concluding chapter of the book which Geddes wrote, just before his death, with J. Arthur Thompson (Geddes and Thompson, 1931).

[5] One of the fascinations of re-reading Geddes is to see again the way that his 'thinking machine' approach leads him to coin so many conflated terms: eutopia and u-topia, sociography, eutechnics, parasitopolis, nekropolis, patholopolis, tyrannopolis.

[6] Later, in a fascinating account of his first ill-starred meeting with Geddes in 1923, Mumford was to suggest a list of other more analytical books by other authors which had 'left a deeper impression than any single work of Geddes: but the impact of his person shook my life to the core' (Mumford, 1966, 15). The long-delayed meeting was not a success; Mumford has the grace to acknowledge that 'two demanding, self-absorbed egos' were involved.

[7] It is significant that Herbertson's wife wrote the first English biography of Le Play. Written in the 1890s, parts of it appeared in *Sociological Review* from 1920 onwards, and it was fully published only in 1950.

References

Abercrombie, P. 1933: *Town and country planning* (London).
Abrams, P. 1968: *The origins of British sociology: 1834–1914* (Chicago).
Beaver, S. H. 1962: The Le Play Society and fieldwork. *Geography* 47, 225–40.
Bergson, H. L. 1911: *Creative evolution*. English translation by A. Mitchell (London).
Boardman, P. 1978: *The worlds of Patrick Geddes: biologist, town planner, re-educator, peace warrier* (London).
Brooke, M. Z. 1970: *Le Play: engineer and social scientist – the life and work of Frédéric Le Play* (London).
Cherry, G. E. 1974: *The evolution of British town planning: a history of town planning in the United Kingdom during the 20th century and of the Royal Town Planning Institute, 1914–74* (Leighton Buzzard).
Cornish, V. 1923: *The great capitals: an historical geography* (London).

Defries, A. 1927: *The interpreter Geddes: the man and his gospel* (London).

Dickinson, R. E. 1947: *City region and regionalism: a geographical contribution to human ecology* (London).

Dickinson, R. E. 1969: *The makers of modern geography* (London).

Faris, R. E. L. 1967: *Chicago sociology: 1920–1932* (San Francisco).

Fawcett, C. B. 1922: British conurbations in 1921. *Sociological Review* 14, 111–22.

Fleure, H. J. 1953: Patrick Geddes (1854–1932). *Sociological Review* New Series 1, no. 2, 5–13.

Garnett, A. 1969: Obituary (H. J. Fleure). *Geography* 54, 466–9.

Geddes, P. 1881: The classification of statistics and its results. *Proceedings of the Royal Society of Edinburgh* 11, 295–322.

Geddes, P. 1884: An analysis of the principles of economics. *Proceedings of the Royal Society of Edinburgh* 12, 943–80.

Geddes, P. 1904a: *City development: a study of parks, gardens, and culture institutes – a report to the Carnegie Dunfermline Trust* (Edinburgh and Westminster).

Geddes, P. 1904b: Cities as applied sociology. Part I. *Sociological Papers* 1, 103–18, discussion 119–44.

Geddes, P. 1905: Cities as concrete and applied sociology. Part II. *Sociological Papers* 2, 57–111, discussion 112–19.

Geddes, P. 1915: *Cities in evolution: an introduction to the town planning movement and to the study of civics* (London). Abridged version reprinted 1949 (introduction by J. Tyrwhitt). Reprinted 1968 (introduction by P. Johnson-Marshall).

Geddes, P. 1924: A proposed co-ordination of the social sciences. *Sociological Review* 16, 54–65.

Geddes, P. and Thompson, J. A. 1931: *Life: outlines of a general biology*, 2 vols. (London).

Geographical Teacher 1910: (Syllabuses of courses given at the Oxford summer school) *Geographical Teacher* 5, 337–53.

Gilbert, E. W. 1972: *British pioneers in geography* (Newton Abbot).

Glass, R. 1955: Urban sociology in Great Britain: a trend report. *Current Sociology* 4, no. 4.

Halliday, R. J. 1968: The sociological movement, the Sociological Society and the genesis of academic sociology in Britain. *Sociological Review* New Series 16, 377–98.

Herbertson, D. 1950: *The life of Frédéric Le Play*. Edited by V. Branford and A. Farquharson (Ledbury).

Kitchen, P. 1975: *A most unsettling person: an introduction to the ideas and life of Patrick Geddes* (London).

Mairet, P. 1957: *Pioneer of sociology: the life and letters of Patrick Geddes* (London).

Marr, T. R. 1904: *Housing conditions in Manchester and Salford: a report prepared for the Citizens Association for the improvement of the unwholesome dwellings and surroundings of the people* (Manchester and London).

Mellor, H. E. 1973: Patrick Geddes: an analysis of his theory of civics, 1880–1904. *Victorian Studies* 16, 291–315.

Mellor, H. E. 1977: Patrick Geddes as an international prophet of town planning before 1914. Paper to *First International Conference on the History of Urban and Regional Planning* (London).

Mogey, J. 1956: 'La science sociale' in England. *In* Le Play, A., *Recueil d'études sociales à la mémoire de F. Le Play* (Paris) 57–64.

Morgan, C. Lloyd 1923: *Emergent evolution* (London).

Mumford, L. 1934: *Technics and civilization* (London).
Mumford, L. 1956: *The human prospect* (London).
Mumford, L. 1966: The disciple's rebellion. *Encounter*, September 1966, 11–21.
Nossiter, B. D. 1978: *Britain: a future that works* (London).
Park, R. E. and Burgess, E. W. 1924: *Introduction to the science of sociology*, 2nd edition (Chicago).
Russell, B. 1946: *A history of western philosophy and its connection with political and social circumstances from the earliest times to the present day* (London).
Tyrwhitt, J. 1949: Introduction in Geddes, P., *Cities in evolution*. New and revised edition (London).
Wells, A. F. 1935: *The local social survey in Great Britain* (London).

12

Royce's 'Provincialism' A Metaphysician's Social Geography

J. Nicholas Entrikin

The writings of America's foremost absolute idealist philosopher, Josiah Royce (1855–1916), provide an illustration of the similarity of many of the contemporary themes of humanist social geographers and the themes of idealist social philosophers of the nineteenth century. Royce's ethical philosophy was based upon the importance of loyalty to a cause, and one of his favourite causes was that of local community, or province. The province was defined by Royce (1908a, 61) as a social and geographic 'whole', and loyalty to this unit was termed 'provincialism' (Royce, 1908a, 1909). Through his discussions of local community and loyalty to such communities, Royce was able to illustrate the practical application of his otherwise abstract religious and social philosophy. If one concentrates on Royce's secular arguments, the similarity between his thought and contemporary ideas of humanist social geographers becomes evident.

Royce's province is an illustration of the linkage of concepts of social organization and spatial organization or place, a familiar theme in social geography (Buttimer, 1968, 134–45; Herbert, 1973, 18–19; Jones, 1975, 7; Pahl, 1965, 81). This theme has traditionally been expressed by concepts such as 'community', 'milieu' and 'region', but their holistic nature posed definitional problems for social scientists. Holistic concepts do not cohere with prevailing views on scientific concept formation, and as a result such concepts are defined operationally in order to negate 'emergent' attributes such as 'structure' or 'wholeness' (Brodbeck, 1958; Stoddart, 1967, 520). Often problems associated with the ambiguity of community, milieu or region have been circumvented by the adoption of terms such as 'action space', 'social space' and 'social area' which refer to distinct areal units whose bounds are determined by the occurrence of a social variable or set of social variables.

This analytic tendency among geographers imbued with the methods and philosophy of positivist social science has been criticized recently by geographers of a variety of orientations (for example, Relph, 1970, 197–9; Harvey, 1973, 286–314). Geographers of one such orientation, humanism, have argued that positivism provided an inappropriate philosophical context for the study of the relationship of man, place and society (Buttimer, 1974, 1976; Ley, 1977; Ley and Samuels, 1978; Relph, 1970, 1976; Tuan, 1971, 1976). They have argued that phenomenological and existential philosophies provide more appropriate contexts in that such orientations are synthetic rather than analytic and thus their practitioners do not separate subject from object or fact from value. Community, milieu and region are important concepts for such geographers in that humanists have argued that one must study the context of behaviour in order to understand human actions. This understanding is only possible when one gains insight into the 'richness' of meaning of these contexts, and such insight demands a holistic perspective.

In tracing the lineage of their approach, humanists have referred often to the Vidalian school and, to a lesser extent, to Park's Chicago school of urban sociology (for example, Buttimer, 1968, 136; 1976, 283; Ley, 1977, 499–503). Both groups displayed some concern for the humanistic study of community, milieu and region, yet, according to Ley:

> In each instance, a humanistic perspective with a holism incorporating subject and object was compromised for a materialist treatment drawing upon a physical science tradition and encouraging deterministic thinking. Social relations were suppressed in the interest of spatial facts, and social geography became preoccupied with man's material works and the irresistible objectivity of the map. (Ley, 1977, 500)

Ley argued that this latter approach, characterized as materialistic, objective and deterministic, has been the dominant approach within social geography, and that social geographers should recapture the more humanistic elements of Vidalian regional geography and Park's human ecology. This humanistic orientation was 'one concerned with holism, a man-environment dialectic, and the incorporation of social and cognitive variables' (Ley, 1977, 502).

Ley's statements were presented in his article in support of the contemporary significance of Alfred Schutz' (1962) phenomenological sociology, but these same statements may be used to describe idealist approaches to the study of society. This similarity is not surprising in that the philosophy of the acknowledged founder of the phenomenological movement, Edmund Husserl, has been described as a form of 'transcendental idealism' (Kockelmans, 1967, 193). Also, two of the most frequently referred to antecedents of the humanist movement, Robert Park and Vidal de la Blache, were well versed in the literature of neo-Kantian idealism. Park (1972, 3) received his PhD under the direction of neo-Kantian philosopher Wilhelm Windelband, and Berdoulay (1976) has noted the influence of French neo-Kantians on the regional geography of Vidal and his students.

Royce's philosophy will be used as a vehicle for illustrating some of the common themes of idealists and humanist social geographers. He was concerned with establishing the relevance of idealist philosophy for examining the major issues of his times. Thus, unlike most idealists, he discussed scientific developments within the context of idealism, as well as discussing the more traditional idealist subjects of morals and religion. Also, Royce had much to say about the social problems of his period, such as the diminished role of local community in American society.

Royce's consideration of the weakened significance of local community in an increasingly urban and industrial society has led intellectual historians and sociologists to identify him as one of the leading communitarians of the early twentieth century (Quant, 1970, 3–4; Wilson, 1968, 27). He extolled the virtues of the small local community and criticized the trend toward increased centralization in American society (Royce, 1908a, 57–96). As a result of his statements on this topic, Royce has been labelled by Lucia and Morton White (1962, 180–3) as one of the American 'intellectuals against the city'.

Geographers have rarely referred to Royce's work, although he frequently discussed geographic topics (Brigham, 1903, 295). His discussions of the physical geography of the Pacific coast were quite descriptive and often impressionistic, and thus attracted little attention from professional geographers. The more interesting aspect of Royce's work, especially for social geographers, was his discussion of the local community or province which he found exemplified in California in the last half of the nineteenth century. Before presenting Royce's discussion of these topics, I will introduce the intellectual context of Royce's work and provide a synopsis of some of the major elements of his philosophy.

ROYCE'S INTELLECTUAL CONTEXT

Royce was a philosopher at Harvard during the so-called 'classical age' of the Harvard philosophy department. He was one of the major figures in the department along with William James, Hugo Münsterberg, G. H. Palmer and George Santayana (Kuklick, 1977). The absolute idealism of Royce was in marked opposition to the very popular pragmatism of his colleague William James, a situation which was painfully obvious to Royce (Pomeroy, 1971, 13–14). This awareness, however, was an important factor in explaining Royce's tendency to emphasize the applicability of his philosophy to the social problems of his day and thus develop a 'public philosophy' (Kuklick, 1977, 291–314).

Royce's philosophy was criticized as an Hegelian idealism which was imported to the United States without consideration of its inappropriateness for the American context (Mead, 1929–30, 211–31; Perry, 1938, 5–25). His supporters argued, however, that Royce's work represented a unique blend of three primary influences: (1) his childhood and education in frontier California (Robinson, 1968, 17–32), (2) his exposure to American Puritanism (McDermott, 1969, 5), and (3) his studies with German idealists Herman Lotze, Wilhelm Wundt and Wilhelm Windelband (Buranelli, 1964, 48–65). According to Royce his first lessons concerning the nature of community occurred during his youth in the mining town of Grass Valley, California and his early school days in San Francisco. Royce wrote of these lessons stating that:

My comrades very generally found me disagreeably striking in my appearance, by reason of the fact that I was redheaded, freckled, countrified, quaint, and unable to play boys' games. The boys in question gave me my first introduction to the 'majesty of community'. The introduction was impressively disciplinary and persistent. (Royce, 1916a, 126–7).

He received his bachelor's degree at the newly founded University of California in 1875. After studying in Germany and receiving his PhD in philosophy at Johns Hopkins University, Royce returned to Berkeley as an instructor in English literature (Kuklick, 1972, 7). As an ambitious young philosopher, Royce found the Californian intellectual life intolerable. He wrote to William James at Harvard complaining of the poor 'metaphysical climate' of California, and stating that he would return to the East for short periods to restore his 'spiritual health' (Royce, 1878a).

While teaching in California, Royce had little good to say about his fellow Californians. In a letter to his friend, Baltimore businessman George Coale, Royce wrote:

The aims of Californians are like the Coast Range hills in the regions where for the most part, our immense wheat crops are raised. They are mean not lofty, and left to themselves, rather barren. (Royce, 1878b)

Once Royce left California in 1882, to spend the remainder of his career teaching at Harvard until his death in 1916, he developed a more favourable attitude toward the state and its people. This new attitude in its excesses could be described as a romantic vision, but in many of his essays he maintained a critical attitude toward what he perceived as spiritual weakness among the state's population. While at Harvard he wrote a number of papers and books on California, which ranged over a variety of topics from geography to literature. The breadth of his interest can be illustrated by the two books he wrote in which California played a major part. The first was an historical work entitled *California, from the conquest in 1846 to the Second Vigilance Committee in San Francisco: A study of American character* (1886), and the second was a novel entitled, *The feud of Oakcreek: a novel of California life* (1887).

As a child and young adult Royce was trained within the Puritan tradition (S. Royce, 1932). McDermott has argued that Royce belongs in the category of 'distinctive American philosophers' along with William James, Ralph Waldo Emerson and Puritan divine Jonathan Edwards, and that Royce's thought resembles that of Edwards (McDermott, 1969, 5n). One of the noted similarities between Royce's thought and Puritan doctrine was the tendency to begin with the experience of suffering and submit 'this experience to intense and complex reflection, thereby hoping to build a new community, a Zion through which the Lord would show His presence' (McDermott, 1969, 5n). Royce followed a similar format in his discussion of community which was presented as a response to the problem of evil in the world (McDermott, 1969, 6).

Royce's studies in Germany provided him with an exposure to German idealism, both Hegelianism and neo-Kantianism (Buranelli, 1964, 57–65). With Lotze, Royce studied a system of idealist philosophy which recognized the necessity of incorporating the advances of scientific thought. Royce's studies with Windelband were primarily concerned with the history of philosophy, and from Wundt, he learned of the significance of the 'collective consciousness'. This interest in a social consciousness, which was later stimulated by Royce's reading of work by American pragmatist philosopher Charles Peirce, was an influential antecedent to his communitarian philosophy (Smith, 1950, 19–33).

Although Royce's work was primarily in philosophy and most of his intellectual contacts were in that field, he also had contacts with a number of geographers and other academics with geographical interests. Geographer and historian Daniel Coit Gilman, a friend and benefactor, followed closely developments in Royce's academic career. Gilman was president of two universities which Royce attended, the University of California and Johns Hopkins (Clendenning, 1970: Franklin, 1910). Royce was a frequent visitor to the Gilman household while he was a student at Johns Hopkins and a correspondent while he was teaching at the University of California at Berkeley. One can only speculate on the topics of their private conversations, but their correspondence indicated that Gilman's role was primarily one of college president, rather than one of geographer (Clendenning, 1970).

Another geographer whose name appeared in Royce's papers was William Morris Davis. Unlike Royce's relationship with Gilman, the Royce-Davis relationship was primarily professional[1] and records of their contacts indicated that their discussions centred upon geography, geology and philosophy (Costello, 1963, 106, 109). Royce had been one of the few absolute idealists to be concerned with accommodating science and idealism and he used Davis' work, among others, to illustrate some of his arguments (Royce, 1915–16, 100). Royce held a seminar described as 'A comparative study of the various types of scientific method' and Davis was an occasional guest and speaker at this seminar[2] (Costello, 1963, 106n). The philosopher was especially interested in the scientific method as used by Davis in his discussions of the evolution of landforms (Royce, 1915–16, 100).

Historian Frederick Jackson Turner met Royce at Harvard and favourably cited his work on provincialism or loyalty to local community (Turner, 1920, 157–8; Pomeroy, 1971, 6). Turner stated that:

> In a notable essay Professor Josiah Royce has asserted the salutary influence of a highly organized provincial life in order to counteract certain evils arising from the tremendous development of nationalism in our own day . . . Whatever may be thought of this philosopher's appeal for a revival of sectionalism, on a higher level, in order to check the tendencies to a deadening uniformity of national consolidation (and to me this appeal, under the limitations which he gives it, seems warranted by the conditions) – it is certainly true that in the history of the United States sectionalism holds a place too little recognized by historians.
>
> . . . Within this vast empire there are geographic provinces, separate in physical conditions, into which American colonization has flowed and in each of which a special society has developed with an economic, political and social life of its own. (Turner, 1920, 1957–8)

This latter statement corresponded to Royce's description of provinces and especially the province of California.

Sociologist Robert Park studied philosophy at Harvard before turning to sociology and founding the Chicago School of urban sociology (Hughes, 1968). Although Park was most influenced by William James during this period, he also took classes from Royce (Baker, 1973, 255–56). No direct evidence is available, however, to support the interesting speculation that Park's concern for community was influenced by his contact with Royce. A number of similarities existed in their respective discussions of community, the two most notable being their mutual interest in the social control aspects of community and their concern for maintaining individual freedom within a strong community structure.

A final intellectual tie of Royce's was with University of California geologist Joseph le Conte. As an undergraduate at Berkeley, Royce took six geology courses from Le Conte which accounted for most of Royce's scientific training (Clendenning, 1970, 15–16).[3] Le Conte was also acknowledged by Royce as one of his principal philosophical influences, for Le Conte was a scientist who viewed the naturalist conception of the universe as the scientist's 'working clothes' and argued that the idealist conception was a more appropriate one (Le Conte, 1908, 301–03). For Le Conte, '. . . all truth is the image of God in human reason' (Le Conte, 1903, 336).

The teleological perspective of Le Conte was similar to that of his friend, geographer Arnold Guyot,[4] and was a philosophy of science which Royce supported (Le Conte, 1903, 146–53). The spirit of this philosophy was expressed in a passage from Guyot's work as cited by Stoddart:

> Few subjects seem more worthy to occupy thoughtful minds than the contemplation of the great harmonies of nature and history. The spectacle of the good and the beautiful in nature reflecting every-where the idea of the Creator, calms and refreshes the soul . . . Every being, every individual, necessarily forms a part of a greater organism than itself, out of which one cannot conceive its existence, and in which it has a special part to act . . . (Guyot, 1850, 85; Stoddart, 1967, 515–16)

This reconciliation of science and religion was an aspect of Le Conte's thought and of Guyot's, which Royce maintained in his philosophy.

ELEMENTS OF ROYCE'S PHILOSOPHY

Royce's philosophy was that of absolute idealism (Werkmeister, 1949, 133–68). Idealism is usually associated with the argument that the mind or spirit is an active agent in our experience, and that our knowledge of the world is based upon ideas about the world rather than on direct perceptions of an external reality (Barrett, 1932). The adjective 'absolute' refers to the postulated existence of a supra-personal mind or spirit. Often such absolutes are equivalent to ideas of god as an all-knowing omniscient observer. In Royce's philosophy, however, community became the absolute. Royce recognized this evolution when he stated that:

I strongly feel that my deepest motives and problems have
centred upon the idea of community, although the idea has only
come gradually to my consciousness. (Royce, 1916a, 1929)

Philosopher John Smith described this evolution when he argued that:

Speaking generally, the all-important change in Royce's
conception of the Absolute consists in the shift from the idea of
consciousness apprehending at a glance all truth and harmonizing
at once all conflicts between the multiplicity of finite wills in
existence, to the idea that the Infinite is actual as a well-ordered
system (or ultimately, community) having a general triadic form and
involving a type of cognition called interpretation. (Smith, 1950, 5)

The triadic form of the absolute consisted of God, the body of Christ or the
Church and the individual. As Kuklick argued:

From the point of view of Royce's metaphysics the 'love of Christ'
is the third or mediating idea which creates the community and
interprets God to finite individuals; Christ is the spirit of the
community who reconciles the finite with the infinite. (Kuklick,
1972, 228)

The 'community' was not limited to only Christians, however, but rather
included 'all mankind' (Royce, 1968, 195).

The triadic form of the community of interpretation pervaded Royce's
entire philosophy, not only as the structure of the absolute, but also as the
basis for the individual's knowledge of self and knowledge of the external
world. The origin of this idea of community of interpretation was
attributed by Royce to American pragmatist philosopher Charles Peirce
(Royce, 1968, 273–95). For Royce and Peirce such communities had a
triadic structure in which (1) an individual (interpreter) interprets (2) an
object (or person) to (3) another person (Royce, 1968, 287). This same
process can be applied to self-reflection in which an individual (present)
interprets a past self in order to plan the actions of a future self. Thus the
structure of interpretation is similar to the triadic structure of time (past,
present, future) and the process of interpretation of self is the same
process as the scientific interpretation of any evolutionary process. For
example, Royce stated:

In fact what our own inner reflection exemplifies is outwardly
embodied in the whole world's history . . . And so, wherever the
world's processes are recorded, wherever the records are preserved, and wherever they influence in any way the future course of

events, we may say that (at least in these parts of the world) the present potentially interprets the past to the future, and continued to do so *ad infinitum*.

Such, for instance, is the case when one studies the crust of a planet. The erosions and the deposits of a present geological period lay down traces which if read by a geologist, would interpret the past history of the planet's crust to the observer of future geological periods.

Thus the Colorado Cañon, in its present condition, is a geological section produced by a recent stream. Its walls record, in their stratification, a vast series of long-past changes. The geologist of the present may read these traces, and may interpret them for future geologists of our own age. But the present state of the Colorado Cañon, which will ere long pass away as the walls crumble, and as the continents rise or sink, will leave traces that may be used at some future time to interpret these now present conditions of the earth's crust to some still more advanced future, which will come to exist after yet other geological periods have passed away.

In sum, if we view the world as everywhere and always recording its own history, by processes of aging and weathering, or of evolution, or of stellar and nebular clusterings and streamings, we can simply define the time order, and its three regions – past, present, future, – as an order of possible interpretation. That is, we can define the present as, potentially, the interpretation of the past to the future. The triadic structure of our interpretations is strictly analogous, both to the psychological and to the metaphysical structure of the world of time. And each of these structures can be stated in terms of the other. (Royce, 1968, 288–9)

Thus the community of interpretation not only provided the form of Royce's absolute, but also represented the basic unit which structures our experience of the world. Knowledge of reality and truth were derived from this process of interpretation. For Royce and Peirce knowledge was an act of interpretation of signs (Smith, 1950, 20–7). Intuition, perception and cognition were incomplete descriptions of the process of knowing in that interpretation was ignored.

Royce's premier example of the community of interpretation was the scientific community. Following the arguments of Peirce, Royce (1915–16, 46) stated that the 'scientific process' could be divided into three stages: (1) 'retroduction' or 'abduction', (2) 'deduction' and (3) 'induction proper'. These stages corresponded to a scientific method in which the scientist (1) collected information concerning a 'trial generalization' or hypothesis, (2) formalized the hypothesis and deduced consequences from it and (3) tested the hypothesis by verifying one of its consequences (Royce, 1915–16, 47–8). Although the individual scientist may make a discovery, his work was not scientific until it was confirmed by others in the scientific community (Royce, 1968, 321–7). One of the illustrations Royce used in discussing this process was the work of W. M. Davis on the evolution of the Connecticut Valley (Royce, 1915–16, 100).

The scientific community of interpretation provided the model for Royce's discussion of how individuals gain knowledge. All knowledge was gained through society. For example, our experience of nature was determined by categories established through social intercourse. Royce stated that:

> . . . social consciousness gives us the only notion of finite reality that we can have . . . a phenomena of nature . . . is solely something suggested to us by agreement between the series of experiences present in various men. And no purely physical experience can possibly prove that nature has other reality than this, viz., reality as a series of parallel trains of experience in various people. (Royce, 1969, 448)

Also, man's knowledge of self was socially defined (Cotton, 1954). Royce rejected arguments that knowledge of self was gained through intuition and instead maintained that the individual ego and the social consciousness were inseparable, each growing with the other. He argued that:

> Just as there is no conscious Egoism without some distinctly social reference, so there is, on the whole, in us men, no self-consciousness apart from some more or less derived form of the social consciousness. I am I in relation to some sort of non-ego. (Royce, 1969, 427)

Individualism was thus a socially created mode of being, which was the result of a well developed social system (Royce, 1968, 113). Although individualism led to tensions within communities, and was a potentially disruptive force, it also provided a source of strength to the community. This strength derived from the loyal bond of independent individuals united by a common purpose, rather than the fusion of many individuals into a single way of thinking within which they lost their independence of thought. According to Royce (1908a, 81–96; 1916b, 238–48) groups which he considered to be a threat to the community order, such as unions, mass movements and mobs, were characterized by this latter tendency.

The loyal bond linking the individual to a large social whole was the foundation of Royce's ethical philosophy. The basic tenet of his ethics was that of 'loyalty to loyalty' (Royce, 1916b; Fuss, 1965). If an individual lived by this principle, he could never be loyal to purely selfish interests or to the selfish interests of another individual. Rather, he would be always loyal to a cause. Such loyalty served to unite individuals within a community. For example, the scientific community was united by the loyalty of its members to the cause of truth (Smith, 1950, 5).

One of the examples which Royce used to illustrate the application of his ethical philosophy was that of the loyalty to local community, or 'provincialism' (Royce, 1908a, 57–108; 1909; 1916b, 245–8). The province as defined by Royce had an elastic quality which allowed it to be applied to areas of varying sizes from 'small towns' to 'great cities' and from 'counties' to 'large sections of the country' (Royce, 1908a, 57–67). The critical characteristic which unified provinces was a sense of 'wholeness' as indicated by the following statement by Royce:

> . . . a province shall mean any one part of a national domain, which is, geographically and socially, sufficiently unified to have a true consciousness of its own unity, to feel a pride in its own ideals and customs, and to possess a sense of its distinction from other parts of the country. And by the term 'provincialism', I shall mean, first, the tendency of such a province to possess its own customs and ideals; secondly, the totality of these customs and ideals themselves; and thirdly, the love and pride which leads the inhabitants of a province to cherish as their own these traditions, beliefs and aspirations. (Royce, 1908a, 61)

The local community was only one of a number of different types of community to which an individual belonged. Thus a person has loyalties to his local community, his state, his country and finally to the community of mankind. Just as a single community was composed of a group of individuals, so a nation was composed of individual communities or provinces, and just as a community gained strength from the independence of its members, so the nation gained strength from the independence of its provinces. Such independence must always be tempered, however, by a spirit of 'wise provincialism' or else a dangerous sectionalism could develop as witnessed in the United States prior to the Civil War and in Australia in the 1880s (Royce, 1908a, 62–7; 1889, 1813–20).

Loyalty to the province was not an abstract ideal, according to Royce, but was rather a 'practical guide for action' which provided important training for loyalty to the nation and loyalty to the world community. Such training was critical for combatting the 'levelling tendencies' which Royce saw at work in American society, tendencies such as (1) the increased mobility of the population, (2) the development of a mass media, (3) the centralization of industry and (4) the growth of a mob spirit (Royce, 1908a, 57–96). These tendencies were viewed as contributing to a tyranny of the social order which destroyed the integrity of provincial customs and provincial independence. Royce's view of the events of his period was to see change in American society as causing a degeneration of moral order and he retreated to the conservative stance of extolling the virtues of the small community, similar to many other intellectuals of his period (Quant, 1970, 137–57). For Royce, local community represented an enlightened form of social control based upon free will and loyalty of its members, as opposed to the repressive control of a national bureaucracy or of a large corporation.

Royce's description of provincialism as a practical guide for action was illustrated by his involvement in a debate over the merits of a centralized plan for higher education in the state of Vermont. In 1914 the Carnegie Foundation for the Advancement of Teaching published a report which suggested a centralization and standardization of higher education in the state of Vermont (Royce, 1914; 1915). On two separate occasions Royce criticized the report of the Foundation and argued that universities and colleges were important institutions for maintaining the distinctiveness of the province. He wrote that:

> When you undertake to plan for the wise guidance of a special province of this country, plan in any case to have your province develop for itself in accordance with its own traditions and with a full sense that every state and province of this country *best* serves the whole country when it grows in accordance with its own soul and in the light of its own experience of what is best in its past, and in its own spirit, in its traditions, in the consciousness and patriotism of its villages, of its country population and of its best days. (Royce, 1914, 11)

Provincialism worked for the benefit of the individual as well as the society as a whole (Brown, 1950, 6–7). At the individual level the maintenance of provincial spirit fights the growing sense of 'self-estrangement', a concept that Royce borrowed from Hegel and identified as a growing ailment of his times (Royce, 1916b, 238–44). This estrangement or alienation produced by impersonal social forces demoralized the individual and thus affected the community. Provincialism not only combatted such alienation, but also contributed to a stronger national society. Royce argued this point when he stated:

> Keep the province awake, that the nation may be saved from disastrous hypnotic slumber so characteristic of the excited masses of mankind. (Royce, 1908a, 96)

H

Royce wrote much of his geographical work in connection with the topic of provincialism. The province which Royce discussed most often was his native California, a region of the United States which he saw as a unified 'whole'.

ROYCE'S CALIFORNIA

California, according to Royce, was a province, composed of many smaller provincial units, and part of a larger whole, the nation. The history of the Anglo-settlement of California illustrated a series of stages which Royce (1909, 234–36) considered common to all newly formed communities: (1) an initial peaceful order based upon the residue of social constraints from the former communities of the new migrants, (2) a period of disorder as older constraints break down and new community loyalties have yet to form, and (3) a gradual growth of social cohesion as the commitment to local community has begun to develop with a more stable population. This process represented for Royce the success of the forces of order over those of disorder and thus the power of 'great social ideals' (Royce, 1900, 6). The essential element in this evolutionary process was the growth of a provincial consciousness, a type of group consciousness reminiscent of the collective consciousness of Wilhelm Wundt and the community of interpretation of Charles Peirce (Muirhead, 1931, 404). Without such a group consciousness, no social order could exist. Royce illustrated this relationship in stating that:

> Its [California's] disorderly stages were acquired social diseases, due to the lack of any settled community consciousness, due to the absence of loyalty on the part of the individual in his relation to his town and to his State. Because the provincial consciousness was lacking, the community tended to a rapid degeneration into a disorderly state. (Royce, 1909, 235)

He further stated that:

> I have come to know how vital for the very conception and existence of any rational social order a provincial consciousness is. (Royce, 1909, 237)

This consciousness evolved as individuals began to share a common set of past and present experiences, and perceived themselves as sharing a common set of future goals. Thus the provincial consciousness formed in a manner similar to all communities of interpretation. An added feature of the development of a provincial consciousness, however, was the inhabitants' shared experience of the natural environment.

Royce on occasion was prone toward making deterministic statements about the relationship of the natural environment and human behaviour which contradicted his voluntaristic philosophy. For example, George Santayana wrote in a somewhat humorous tone:

> . . . I remember my old teacher, Josiah Royce (who was a native of California), shaking his head at the Middle West, and saying that great plains produce flat and monotonous minds, whereas mountainous coasts, archipelagos and islands are the soil of genius and liberty. (Santayana, 1969, 207)

Although a number of commentators have noted Royce's deterministic tendency (for example, Pomeroy, 1971, 18) a closer inspection of his work reveals a philosophy more appropriately described as environmental possibilism. This possibilistic orientation was best expressed in a paper presented to the National Geographic Society in 1898. In this paper entitled 'The Pacific coast: a psychological study of the relations of climate and civilization', Royce argued that the natural environment, especially the steadiness of climate, had:

> . . . a moral significance in the life of California [that] is of course both good and evil, since man's relations with nature are, in general, a neutral relation upon which ethical relationships may be based. If you are industrious, this intimacy with nature means constant cooperation, a cooperation never interrupted by frozen ground and deep snow. If you tend to idleness, nature's kindliness may make you more indolent. . . Moreover, the nature that is so uniform also suggests in a very dignified way a regularity of existence, a definite reward for a definitely planned deed. (Royce, 1908b, 208–9)

The environment thus played an important role in the development of a provincial consciousness, but it eliminated neither free will nor individual differences. Rather, the strength of the California community was derived from the independent spirit of its population. Royce noted that this spirit of its members was transferred to the spirit of the province as a whole when he stated that:

> During these years [1850–6] many who had come to Calfornia without any permanent purpose decided to become members of the community, and decided in consequence to create a community of which it was worthwhile to be a member. The consequence was the increase of the influence of the factor of geographical isolation in its social influence upon the life of California. The community became self-conscious, independent, indisposed to take advice from

without, very confident of the future of the state, and of the boundless prosperity soon to be expected; and within the years between 1860 and 1879 a definite local tradition of California life was developed upon the basis of the memories and characters that had been formed in the early days. The consequence was a provincial California . . .(Royce, 1908b, 212)

The 'wholeness' of provincial California was thus a spiritual unity, derived from the social consciousness. Such a consciousness allowed the individual to define himself through interpretation and comparison, and also provided the categories for his knowledge of the world beyond the self. A person was free to choose to defy the social order through selfish behaviour, yet he never abandoned this order. As Royce stated in summarizing his study of the history of California:

> After all, however, our lesson is an old and simple one. It is the State, the Social Order, that is divine. We are all but dust, save as this social order gives us life. When we think it our instrument, our plaything, and make our private fortunes the one object, then this social order rapidly becomes vile to us; we call it sordid, degraded, corrupt, unspiritual, and ask how we may escape from it forever. But if we turn again and serve the social order, and not merely ourselves, we soon find that what we are serving is simply our own highest spiritual destiny in bodily form. It is never truly sordid or corrupt or unspiritual; it is only we that are so when we neglect our duty. (Royce, 1886, 501)

The natural environment of California was important in the development of a provincial consciousness which served the social order, but in the final analysis the environment derived its meaning, or its 'spirit', through the social consciousness.

CONCLUSION

Royce's concern with community and provincialism has been interpreted as a conservative reaction to the changes American society associated with population growth and the increased urbanization of that population (Wilson, 1968, 164). His ideology was conservative in that he attempted to fight the trend toward centralization by advocating a return to the past in which the local community dominated American life. According to historian Robert Wiebe (1967, xiii), however, the demise of the local community as the focus of American social and political life was an established fact by the 1870s. Royce was thus romanticizing an American collective past as well as his own past in California, but he was not alone in his retreat to the ideal local community as a means for fighting increasing alienation and maintaining social order (Quant, 1970; Wilson, 1968; White and White, 1962).

An interesting aspect of Royce's thought not always shared by others advocating the benefits of local community was that Royce had a clearly specified metaphysical system which provided a coherent context for his philosophy of community. His philosophy was idealist and thus holistic. The community of interpretation provided the basis for Royce's holistic perspective, and the province was the concrete manifestation of such a community. Royce postulated the existence of a group consciousness or provincial spirit which served to unify these provinces.

The postulated existence of a group consciousness or spirit would be unacceptable to most contemporary geographers as would much of the religious orientation of Royce's philosophy. In his work, however, can be found many of the topics which concern contemporary social geographers, such as the social construction of reality, the significance of local community, linkages of man to place and man's relationship to the physical environment. If the reader sifts out much of Royce's religious philosophy from his discussion of provincialism, an orientation similar to that of contemporary humanist geographers emerges. This similarity has been indirectly noted by existentialist philosopher Gabriel Marcel who stated that 'Royce's philosophy – and this is its greatest value – marks a kind of transition between absolute idealism and existentialist thought' (Marcel, 1956, XII).

Royce's voluntarism, or emphasis of the individual will, in his discussions of community and provincialism illustrates the similarity between his philosophy and existentialism. The individual was part of a community, however, and the dialectical relationship of the individual will and the communal will was an important theme in Royce's idealism. The relationship of place and society was in part manifested in the development of this communal will or provincial spirit. Royce's statements concerning this provincialism provide an example of the compatibility of idealism and concepts such as community, milieu and region, in that these terms maintain their sense of 'wholeness' within such a philosophical context. How the humanist social geographers will be able to retain this 'wholeness' and avoid the pitfall of idealism is an issue they have yet to resolve.

Notes

1. They were also neighbours (Chorley et al, 1973, 235).
2. Royce used Davis' article 'The disciplinary value of geography,' printed in the *Popular Science Monthly* 78, 105–119, 223–40, as reading material for his 'methods' seminar (Royce, 1911).

3.   Royce's use of geological and geographical examples to illustrate his philo-
     sophy, and his descriptions of the physical geography of California, can be
     attributed to the influence of Le Conte's courses. The descriptions of
     Le Conte's geology courses as found in the University of California registers
     for 1871–5 indicated that they covered three topics, (1) 'Dynamic Geology
     . . . the study of the various agencies now at work modifying the earth's
     surface and producing structure', (2) 'Structural Geology' and (3) 'History
     of the Earth'. The historical aspect of the course was to be illustrated 'princi-
     pally from American geology and, as far as possible, from the geology of the
     state' (University of California 1874, 39–40).
4.   Guyot was also influential in Gilman's development as a geographer. (See
     Franklin, 1910, 15; Wright, 1961.)

References

Baker, P. 1973: The life histories of W. I. Thomas and Robert Park. *American Journal of Sociology* 79, 243–60.
Barrett, C. 1932: Introduction. In Barrett, C., editor, *Contemporary idealism in America* (New York) 13–21.
Berdoulay, V. 1976: French possibilism as a form of neo-Kantian philosophy. *Proceedings of the Association of American Geographers* 8, 176–9.
Brigham, A. 1903: *Geographic influences in American history* (Boston).
Brodbeck, M. 1958: Methodological individualisms: definition and reduction. *Philosophy of Science* 25, 1–22.
Brown, S. G. 1950: *The social philosophy of Josiah Royce* (Syracuse, New York).
Buranelli, V. 1964: *Josiah Royce* (New York).
Buttimer, A. 1968: Social geography. In Sills, D., editor, *International encyclopedia of the social sciences* 6, 134–45.
Buttimer, A. 1974: Values in geography. *Association of American Geographers Resource Paper*, 24.
Buttimer, A. 1976: Grasping the dynamism of lifeworld. *Annals of the Association of American Geographers* 66, 277–92.
Chorley, R., Beckinsale, R. and Dunn, A., 1973: *The history of the study of landforms or the development of geomorphology*, Vol. 2 (London).
Clendenning, J. J. 1970: *The letters of Josiah Royce* (Chicago).
Costello, H. 1963: *Josiah Royce's seminar, 1913–14* (New Brunswick, New Jersey).
Cotton, J. H. 1954: *Royce on the human self* (Cambridge, Mass.).
Franklin, F. 1910: *The life of Daniel Coit Gilman* (New York).
Fuss, P. 1965: *The moral philosophy of Josiah Royce* (Cambridge, Mass.).
Guyot, A. 1850: *The earth and man: lectures on comparative physical geography* (London).
Harvey, D. 1973: *Social justice and the city* (Baltimore).
Herbert, D. 1973: *Urban geography* (New York).
Hughes, H. 1968: Robert Park. In Sills, D., editor, *International encyclopedia of the social sciences* (New York) 416–19.
Jones, E. 1975: Introduction. In Jones, E., editor, *Readings in social geography* (London) 1–12.
Kockelmans, J. 1967: Transcendental idealism. In Kockelmans, J., editor, *Phenomenology* (Garden City, New York) 183–93.
Kuklick, B. 1972: *Josiah Royce: an intellectual biography* (New York).

Kuklick, B. 1977: *The rise of American philosophy – Cambridge, Massachusetts 1860–1930* (New Haven).

Le Conte, J. 1903: *The autobiography of Joseph Le Conte* (New York).

Le Conte, J. 1908: *Evolution* (New York).

Ley, D. 1977: Social geography and the taken-for-granted world. *Transactions of the Institute of British Geographers*, New Series 2, 498–512.

Ley, D. and Samuels, M., editors, 1978: *Humanistic geography* (Chicago).

Marcel, G. 1956: *Royce's metaphysics* (Chicago), translated by V. Ringer and G. Ringer.

McDermott, J. J. 1969: Suffering, reflection, and community: the philosophy of Josiah Royce. In McDermott, J. J., editor, *The Basic Writings of Josiah Royce* 1 (Chicago) 3–18.

Mead, G. H. 1929–30: The philosophies of Royce, James and Dewey in their American setting. *International Journal of Ethics* 40, 211–13.

Muirhead, J. 1931: *The Platonic tradition in Anglo-Saxon philosophy* (London).

Pahl, R. E. 1965: Trends in social geography. In Chorley, R. J. and Haggett, P., editors, *Frontiers in geographical teaching* (London) 81–100.

Park, R. 1972: *The crowd and the public* (Chicago).

Perry, R. B. 1938: *In the spirit of William James* (New Haven).

Pomeroy, E. 1971: Josiah Royce, historian in quest of community. *Pacific Historical Review* 40, 1–20.

Quant, J. 1970: *From the small town to the great community* (New Brunswick, New Jersey).

Relph, E. 1970: An inquiry into the relations between phenomenology and geography. *Canadian Geographer* 14, 193–201.

Relph, E. 1976: *Place and Placelessness* (London).

Robinson, D. 1968: *Royce and Hocking, American idealists* (Boston).

Royce, J. 1878a: Letter to William James 16 July. In Clendenning, J., editor, 1970: *The letters of Josiah Royce* (Chicago) 59.

Royce, J. 1878b: Letter to George Coale 16 September. In Clendenning, J., editor, 1970: *The letters of Josiah Royce* (Chicago) 62.

Royce, J. 1886: *California, from the conquest in 1846 to the second vigilance committee in San Francisco: a study in American character* (Boston).

Royce, J. 1887: *The feud of Oakfield Creek; a novel of California life* (Boston).

Royce, J. 1889: Reflections after a wandering life in Australia. *Atlantic Monthly* 63, 675–86, 813–28.

Royce, J. c.1900: The opening of the great West, Oregon and California. Lecture notes contained in the Royce papers, Widener Library, Harvard University Archives, Cambridge, Mass.

Royce, J. 1908a: Provincialism. In Royce, J., editor, *Race questions, provincialism and other American problems* (New York) 57–108. Originally presented as a Phi Beta Kappa address at Iowa State University in 1902.

Royce, J. 1908b: The Pacific coast, a psychological study of the relations of climate and civilization. In Royce, J. editor, *Race questions, provincialism and other American problems* (New York) 169–225. Originally presented as an address to National Geographic Society in 1898.

Royce, J. 1909: Provincialism based upon a study of early conditions in California. *Putnam's Magazine* 7, 232–40.

Royce, J. 1911: Letter to C. Minot 26 March. In Clendenning, J. J., editor, 1970: *The letters of Josiah Royce* (Chicago).

Royce, J. 1914: A plea for provincial independence in education. *Middlebury College Bulletin* 9, 3–19.

Royce, J. 1915: The Carnegie Foundation for the advancement of teaching and the case of Middlebury College. *School and Society* 1, 145–50.

Royce, J. 1915–16: Royce's metaphysics lectures, class notes from Phil. 9, Harvard University, taken by R. W. Brown and B. E. Underwood. In the Royce Papers, Archives of the University Research Library, U.C.L.A., box 2.

Royce, J. 1916a: *The hope of the great community* (New York).

Royce, J. 1916b: *The philosophy of loyalty* (New York).

Royce, J. 1968: *The problem of Christianity* (Chicago). Originally published in 1918.

Royce, J. 1969: Self-consciousness, social consciousness and nature. In McDermott, J. J., editor, *The basic writings of Josiah Royce* (Chicago) 423–61. Originally presented in an address to the Philosophical Club of Brown University in 1895 and later expanded and published in 1898.

Royce, S. 1932: *A frontier lady: recollections of the gold rush and early California* (New Haven).

Santayana, G. 1969: Geography and morals. In Lachs, J. and Lachs, S., editors, *Physical order and moral liberty* (Nashville, Tenn.) 207–09.

Schutz, A. 1962: *Collected Papers*, Natanson, M. editor (The Hague)

Smith, J. 1950: *Royce's social infinite: the community of interpretation* (New York).

Stoddard, D. R. 1967: Organism and ecosystem as geographical models. In Chorley, R. J. and Haggett, P., editors, *Models in geography* (London) 511–58.

Tuan, Y. F. 1971: Geography, phenomenology, and the study of human nature. *Canadian Geographer* 15, 181–92.

Tuan, Y. F. 1976: Humanistic geography. *Annals of the Association of American Geographers* 66, 266–76.

Turner, F. J. 1920: The Ohio Valley in American Society. In Sumner, F. J., editor, *The frontier in American history* (New York) 157–76. Originally presented as an address before the Ohio Valley Historical Association in 1909.

University of California, 1874: *Register*. In the University of California Archives, Powell Library, U.C.L.A.

Werkmeister, W. H. 1949: *A history of philosophical ideas in America* (New York).

White, M. and White, L. 1962: *The intellectual versus the city* (Cambridge, Mass.).

Wiebe, R. 1967: *The search for order, 1877–1920* (New York).

Wilson, R. J. 1968: *In quest of community* (New York).

Wright, J. K. 1961: Daniel Coit Gilman, geographer and historian. *Geographical Review* 51, 381–99.

13

Epistemology and the History of Geographical Thought

Paul Claval

1 EPISTEMOLOGY AND THE PROGRESS OF GEOGRAPHY

(a) The current popularity of epistemology reflects new concern about the validity of scientific method. Throughout the nineteenth century science produced such satisfactory results that to question its certainties seemed unnecessary. Epistemology adopted an inductive approach: it followed science, describing its methods and indicating the procedures needed to establish facts and justify the preferred interpretation of them; subsequently it drew up a conspectus of the principles gradually established by science.

From the beginning of the twentieth century the situation changed: in the fields of mathematics and logic many of the results which had seemed definitively established were called into question. Alongside Euclidean geometry non-Euclidean geometries were invented. Alongside the old Aristotelian logic and the law of the excluded middle, new forms of logic with, for example, multiple values were developed. This type of work, far from remaining purely speculative, proved indispensable to the progress of physical theories. Consequently epistemological research was no longer seen as fundamentally inductive. It was no longer content to follow the movement of science: it undertook to precede science by indicating which tools should be used to make the best progress. This was the atmosphere which characterized neo-positivism: some concern was already apparent, therefore, but epistemology was still only partially critical. Its role was still more prescriptive.

Nowadays there is an awareness of the philosophical shortcomings in the two previous attitudes: objective knowledge cannot be based on itself;

what are the presuppositions? How does it justify its own validity? Epistemological thought is no longer intended to be mainly prescriptive – the sciences establish their own rules – it is critical, it questions the explicit and usually implicit principles behind the approaches taken. Epistemology teaches one to doubt certainties. It illuminates the disciplines from inside and emphasizes how they limit themselves from the start by the presuppositions they accept.

(b) The nineteenth century approach to epistemology is obviously of little interest to the researcher: it appears to consist of a wordy commentary on what he is doing; it is the work of philosophers who have not even taken the trouble to learn the subject and who claim to judge from the outside. It does not and cannot contribute anything and does not always understand wherein lies the real value of the methods used.

Neo-positivist epistemology has quite different qualities: it makes sure that the instruments *a priori* necessary for the formalization of procedures are valid; it is inextricably linked to the new forms of logic and mathematics. It is not merely a codification of well tried methods; it proposes new ways and tests the coherence of the methods already in use. A large proportion of the quantitative revolution and the new geography is modelled on norms laid down in the 1950s by epistemologists for the practice of the social sciences (Harvey, 1969). For example, the exceptionalist point of view which made geography a special case was rejected, and it was conceived on the model of the other social sciences; great account was taken of Karl Popper's ideas (Popper, 1945a, 1945b), his refutation of the historicist points of view which were the basis of exceptionalism at the level of the humanities as a whole was accepted; care was taken in theorizing about class action/practice – this was seen as existing only in the case where the class or group were caught up in the same situational logic; one learned to distrust grand revolutionary ideas and to make the social sciences the privileged tools of *piecemeal social engineering* by which the course of events and the functioning of society could gradually be altered. All this goes hand in hand with the importation of more rigorous techniques than those used previously: progress made by operational research or by systems theory has been so rapid since World War II that there appears to be an inexhaustible reservoir.

Some Marxists also follow prescriptive epistemology (Levy, 1975). They expect it to establish a system of key concepts in the humanities; these have been suggested by the critique of political economy; they only need revamping and a modification of their field for them to be applicable to other human sciences. From this point of view, epistemology is not in control of methods, but dictates the relevant contents; it already contains all universal truth; all that is needed to reveal the truth is to let its potential be developed. This is a reassuring concept since all knowledge is already potentially discovered and only the minor consequences remain to be stated; but it is also discouraging because it reduces present-day humanity, contemporary researchers, to mere successors to a way of thinking which has already overcome all obstacles.

(c) Reflection is much often critical questioning than stated certainty: can one be sure that all the results produced by scientific progress are of equal value? Are there universal criteria available to determine truth? What is the basis of the best constituted branches of learning? In the past 15 years, what used to occupy the thoughts of a few isolated individuals has become a burning question for everyone. Progress usually comes about in a strange way. In nearly all disciplines the idea was that results were increments to a growing store of knowledge; science appeared to be a process of gradual and continuous construction. Then people started to stress the importance of the discontinuities. Before World War II, Gaston Bachelard (Bachelard, 1934-8), who in French-speaking countries holds a position parallel to that of the neo-positivist school in the Germanic and English-speaking worlds, showed that each science arises from an epistemological break; this calls into doubt the evidence of perception and constitutes the objects on which scientific reasoning can attach itself. Gradually, this model of discontinuities was accepted by French philosophers. Althusser gives it authority by applying it to the thoughts of Marx, by showing how, in his opinion, Marx formed the basis of the 'history-continent' of science, i.e. the possibility of humanities, by breaking with the too obvious categories of classical analysis, price, supply, demand, and substituting for them a 'scientific' object, work (Althusser, 1969, 6-7). Marxists are therefore divided between the certainties presented by positive epistemology inspired by the categories of *Capital*, and uncertainty as to whether the epistemological break on which their discipline is founded has yet occurred.

Among the neo-positivists a mutation similar to that caused by Bachelard's thinking in France took place in the 1960s when Thomas Kuhn's work (Kuhn, 1962) on the structure of scientific revolutions was published: all disciplines were thenceforward condemned to ask questions about their past, to trace the discontinuities which made them possible, and to draw up a list of the paradigms on which the disciplines were subsequently based. Kuhn put forward a series of simple notions, easily applicable to all disciplines. The impact of these notions was considerable. He made people aware of the novelty of the currents which had been stirring up geography since the 1950s; he explained the growth of interests in paradigms and even the success of the term 'new geography' proposed by Peter Gould in 1968. In practice however, the instruments of analysis thus popularized turned out to be a little disappointing. What is a paradigm? Do all scientific revolutions take place at the same level? Beyond the discontinuities must one not ask questions about deeper permanences or about directions which, despite all upheavals, maintain the same course on parallel quests?

Foucault-style epistemology (Foucault, 1966, 1969, 1975) belongs to the same group as Bachelardian epistemology: like Althusser's it is based on putting the fundamental gaps and discontinuities into perspective. This type of epistemology puts forward an overall view of the great revolutions in scientific thought, but does so using methods different from those of Thomas Kuhn. Up to now geographers have made little use of Foucault's work, while Foucault himself ignores the discipline of geography.[1]

(d) The history of geographical thought can be seen in two ways: it can be of historical inspiration or of epistemological inspiration. In the first case it attempts to re-create the past, to work out what made geography successful in any particular period, or what linked it to the society in which it was flourishing. In the second case it attempts to work out the logical development of ideas; in order to interpret a movement it does not hesitate to bring together what did not seem to be coherent to people at the time: it makes room for the first economists, for statisticians and for demographers alongside travellers and geographers in the strict sense when it wishes to assess eighteenth-century geographical awareness. It therefore becomes problematic, because it aims not simply to recount the emergence of ideas, but also to explain why what seems to us to spring from the same logic was not recognized as such in the eighteenth century. It therefore questions knowledge by going beyond the statement of facts: it wonders what authority was behind them and penetrates the development of the discipline from the inside.

The history of geographical thought, as it has developed over the past 15 years in France or Germany, for example, has close links with epistemology: the works of Dietrich Bartels (Bartels, 1968) or Gerhard Hard (Hard, 1969, 1970) on the one hand, and those of Josef Schmithüsen (Schmithüsen, 1970, 1976) on the other, take this approach; I wrote my *Essai sur l'évolution de la géographie humaine* as a critical exercise with the intention of demonstrating the logic of different types of human geography and their development in time. Alain Reynaud has studied the coherence of geomorphology, and the mythical background to the geography practised by the average geographer in France in the mid 1960s (Reynaud, 1971, 1974). In the United States Anne Buttimer has studied French geography (Buttimer, 1971). However, the archetype of all these epistemologically oriented studies of the history of thought is the classic work by Richard Hartshorne, *The nature of geography* (Hartshorne, 1939), and also the commentary which he wrote 20 years later, *Perspective on the nature of geography* (Hartshorne, 1959). These studies have greatly contributed towards the spiritual renewal of geography. They are, nevertheless, prisoners of a limited field: they ignore the evolution of thought in parallel disciplines. It is interesting to compare the approach of historians of geographical thought with that of epistemologists whose inspiration is critical: there is undoubtedly no better way of opening up research to new concerns, and at the same time relating it more profoundly to the other social sciences. From this point of view, the work of Michel Foucault can serve as a test.

2 THE EPISTEMOLOGICAL CATEGORIES OF MICHEL FOUCAULT

What are the *idées-forces* of Michel Foucault? We shall consider them as they are presented in *Les mots et les choses* (Foucault, 1966), the work in which he presents a general interpretation of the history of sciences and in particular of humanities, in *L'archéologie du savoir* (Foucault, 1969), in which he justifies his epistemological undertaking, and in *Surveiller et punir* (Foucault, 1975), where his research turns away from sciences to the knowledge which makes them possible and the practices which enclose them. These three works indicate an epistemological reflection on the humanities undertaken between 1955 and 1970, followed by a new orientation in which the author, taken up with the adoption of a new *episteme* and anxious to break with what he considers to be the ambiguous aspects of sciences or humanities, studies them from the new angle of the institutions and behaviours which make their discourse possible.

The positivist epistemologies have been built on the idea that it is possible to express that which escapes direct perception but explains phenomena in depth. Language does not play any role in the theory of knowledge as built up in the nineteenth century, and as we continue to comprehend it through the heritage of positivism and neo-positivism. From this point of view, the value of science can in fact be reduced to the effectiveness of its means of apprehending reality. In order to understand the evolution of scientific thought better, we should stand back from a situation which still impinges too closely. Michel Foucault does so by showing that the relationship between words and things has varied during the course of modern history (Foucault, 1966).

(a) In the sixteenth century both were still perceived at the same level. Words form part of the world; they are its symbols, but at the same time they directly signify the world. Reality is all on one plane; there is no dimension of separation between the world and the discourse which tries to apprehend it. The world reveals itself through words by the interplay of similarities, analogies or identities which, on a real level, relate one object to another in accordance with laws of a universal relationship between things, a sympathy which draws distant things together or a closeness which contaminates them. Explanation is seeking in each symbol the element to which the symbol refers, so that progressively reality is universally transparent and present both in texts and in the world, only undefined gymnastics of interpretation and deciphering are needed.

Sixteenth-century knowledge is therefore determined by the concept of words and things that people accepted at that time: the logic of particular explanations cannot be understood without this analysis of relationships between the world, language and the speaker; scientific discourse has a logic which is circumscribed by the epistemological basis thus defined. To adopt the vocabulary of Michel Foucault, the sixteenth century has a certain *episteme*. This is not, as a summary analysis would suggest, a *Weltanschauung*, a certain conception, a certain colouring of the world characterized by the dominant values, by the common ethical, aesthetic or philosophical preoccupations: it is something deeper and at the same time less conscious, something which depends on the recognized status of the fundamental instruments of knowledge. Compared with the analyses conducted in terms of ideology, the break is also considerable: to understand the nature of thought it must be freed from that which makes it prisoner of the classes or groups of which society consists, it must be assessed through the logic of the instruments which express and condition it.

(b) In the first half of the seventeenth century the Renaissance system of thought was overthrown by the classical system: from then on Western intelligence ceased to perceive things and words at the same level — it established a break between the universe and the representation of it – but it formed an idea of representation which underlay every effort of comprehension at that time. In its view, the symbols refer only to themselves, but at the same time they have the property of giving an account of the ordering of the world; instead of a system on one plane, we have a system on two planes without depth, so that the representation can be so arranged that it superimposes itself perfectly on top of reality. The problem of science is to find a language which truly reflects the ordering of the world – the language of mathematics and mechanics for those in the Galilean and Cartesian tradition concerned with physical realities, but the vernacular for those trying to penetrate the general system of nature. In both cases it is a question of showing up the ordering of reality through the structuring or manufacturing of a language. In this sense the ideal of universal science is embodied particularly well in Adanson's, Tournefort's and Linné's botany: it appears there in the pure linguistic form which elsewhere is beneath the surface.

The classical *episteme* is therefore built on a certain concept of language, its functions and its aptitude at apprehending reality. It does not look for causes in the real world, it does not try to give an account of things through other things; it tries to place them in nature's grand design, or to understand them in terms of what preceded and what followed them, as in a *tableau anime*. The sciences are homologous with general grammar which in one sense is their model; they include an analysis of the proposition, the articulation, the origin and the derivation, which in botany takes the form of an analysis of the structure and the determination of the generic characteristics.

(c) As a result of a profound change in all concepts of knowlege, the classical *episteme* is replaced at the beginning of the nineteenth century by the modern *episteme*. In its movement towards in-depth study at the end of the eighteenth century, science discovered behind analogies of form, organic similarities; it becomes more interested in functions than in appearance and learns to read life's data in sequence. At the same time the study of languages finally passed the stage of general grammar: comparative study of languages led to the formulation of laws about the development of words and grammar, an awareness of the autonomy of aspects of language: they are not transparencies of reality, they are not to be confused with the logic of things, they have a logic of their own. Science is no longer identified with the making of a language, but with the exploration of a reality outside language, of which language, in its dense, imperfectly transparent state, subject to historical development, can give an account only with difficulty. The epistemic basis henceforward brings in not words and things, but the subject observing the things of which he apprehends the links, the structures, the being itself and the underlying forces modelling reality – Foucault points out the sudden attention given to life, work and words. In this context the sciences concerned with man and society suddenly find themselves invested with special status: they are caught in the dialectic of the subject observing who is at the same time the object of study. Foucault shows that such an undertaking is conceivable only if it is thought possible to constitute a science of man not very far removed from philosophical anthropology; he sees there one of the most original elements of the *episteme* which arises in the first half of the nineteenth century.

(d) Analysis of the succession of epistemological bases since the sixteenth century shows up the role in the history of knowledge of discursive practices. *L'archéologie du savoir* (Foucault, 1969) tries to found the study of discursive practices by taking as a model the serial analysis of contemporary history. Instead of reading the history of thought as the history of ideas, their diffusion and their social conditioning, Foucault sees it as an accumulation of discourses. Just as, at the beginning of the nineteenth century, philology perceived the historical nature of language, Foucault shows the historical nature of discursive practices and their succession.

In *Les mots et les choses*, attention was centred exclusively on scientific knowledge, but as soon as one tries to apprehend the status of discourses, what makes them possible, what constitutes their own logic, one is naturally led to moderate the opposition which is usually established between science and other forms of knowledge: discourses modelled on the reality constituting science cannot be constructed without a certain economy of discourse and without certain practices of observation which are also reflected in the knowledge surrounding science. They do not have the same exactness, but they are something quite different from an unstructured mass of precepts or ideologies dictated by interest. *Surveiller et punir* (Foucault, 1975) establishes how the legal and medical sciences were developed at the same time as the development of the detailed techniques of observation which fed them with facts and made them possible.

GEOGRAPHY AND *L'ARCHEOLOGIE DU SAVOIR*

What can the analytical apparatus proposed by Michel Foucault contribute to the analysis of the development of geographical thought? In our opinion a framework for relocating the researches already undertaken in the intellectual environment which sustained them, on the one hand, and a number of questions on the development of modern geography from the end of the eighteenth century on the other.

(a) Epistemological history of thought is often criticized for being too exclusively preoccupied with the questions of our time when trying to shed light on the past, so that the internal logic of development is in danger of eluding the researcher. This is a common fault of those who are anxious to restore the past in all its dimensions: they are often prisoners of contemporary logic and cannot see those qualities of old works which cannot be integrated into our system. Reflections on the *episteme* invite one to plunge into the past using its own eyes. Two examples can be given for this, for the Renaissance and for the classical period.

Brian S. Robinson (1973) has recently published a very curious article on Elizabethan society and its place names. In it he shows how we should understand the logic of everything which appears in the curious compilations of William Camden or Richard Carew who felt obliged, in mentioning Hibernia or Cornwall, to add much irrelevant material which seems to us to bear no relation to our discipline. One finds there a hotch-potch of etymological notes – usually absurd and uninteresting – considerations on the relationships of the names and forms of settlements, and notes on heraldry and myths. Robinson's study is largely inspired by Claude Lévi-Strauss and Edmund Leach's reflections on primitive thoughts and methods of classification – which means that the sources are close to those of Michel Foucault. What Robinson notes in the case of manufacturers of Elizabethan topographies is exactly what, in Foucault's view, is the general characteristic of the thought of the time, the habit of rating equally anything to do with facts and anything to do with names, of putting in the same place natural and physical notes and also notes which belong to folklore, etymology, and history and its symbols. Eva G. R. Taylor's researches into Elizabethan geography have been mainly concerned with topographers and their methods of surveying, with travellers and with teachers. She has unintentionally retained of the geography of the period only the matters with which our discipline is concerned. The jumble of information on place names seems to us to be outside the scientific field. Foucault shows us that at the time it had just as much relevance as what continues to interest us today: it was part of what the *episteme* of the time indicated was knowledge.

A. Downes (1971) was interested in the *Bibliographic dinosaurs of Georgian geography, 1714–1830*. The eighteenth century was the century of encyclopaedias in England, where the movement started, even more so than in France. Among these publications, the great universal geographical compilations have pride of place. They are surprising to us – why such enormous accumulations of facts? Why this endless repetition of the same themes? Why this effort so totally devoid of misgivings and the need to explain facts in a geographical way? Why the lasting commercial success of undertakings which to us seems absurd? Again it is through a lack of intellectual agility and through ignorance of the logic of the time that we cannot get into these works. They corresponded, at a spatial level, to the desire to put things in order, to the concern with drawing up a coherent picture of the universe which led naturalists to write natural histories at the time; the latter noted all the differences enabling minerals, plants and animals to be named and classified, just as the geographers placed on the world map everything worth knowing. Cartographers continued their incessant labelling job and manufacturers of encyclopaedias and bibliographic dinosaurs attached a short note to each place.

Episteme analyses must facilitate studies which, in the image of the two we have just mentioned, situate a work in what provides its logic and necessity. Foucault's schema is useful for understanding the position of the discipline from the Renaissance to the seventeenth or eighteenth century. It shows how the aspects of the geography of former times which have become obsolescent were just as necessary to its implicit presuppositions, just as inevitably inherent in the epistemological base, as the aspects that have survived – for example, the techniques of cartography and the concern for the precise description given by travellers. For the nineteenth century and modern times, Foucault's frameworks are unfortunately less easy to interpret. However, the questions which his discussion poses for geography are capable of better illuminating the development of our discipline – as is the total silence which he maintains on the subject.

(b) The modern *episteme* is characterized by the search, through the exploration of the underlying layers of reality, for causalities which express the fundamental play of forces (Foucault, 1966). Foucault's reasoning depends on the study of three disciplines which seem vital to him: the natural sciences and biology, grammar and linguistics, and political economy. For the three subjects, the principle which enables one to pass from the ordered pictures of the classical age to explanation is the same as that which enables one to see beneath the surface of things the logic of their functioning, and stresses, in their respective fields, the importance of life, work and language; so this principle stresses the finite nature of man, enclosed in each field by a boundary which limits and crushes him – life by the foreknowledge of death, work by the desire which motivates and overtakes it, language by the diversity of what it must express and the imperfection of its means. This finitude gives these disciplines their historicity, leads to natural evolution, economic history and comparative linguistics. Michel Foucault shows very well how all modern thought belongs in the same movement through the awareness of finitude and the idea of the flux of forms and of time.

Foucault's analysis thus has links with Bachelard's and Althusser's – since it stresses the importance of the epistemological breaks which give rise to new sciences by moulding and constructing new objects. We can also see what Foucault owes to Marxism – work to him seems to be the profound structuring reality of all social life. But the schema he suggests is more subtle than that of his predecessors in that it shows both the themes chosen by the new disciplines, life, work, or language, and the privileged path they take – that of historical analysis.

Not all social sciences are equally advantaged by this transformation: those which prosper are those which can easily find a place in an evolutionist framework. Take, for example, political economy: it had studied spatial problems in great detail in the mercantilist period; but from Ricardo onwards it only concerned itself with the development of production through time.

It was not therefore with the social sciences that the new discipline of geography could find a place in the system of knowledge as it was at the time. In the field of life sciences things were different: as Foucault rightly shows, from the time when, with Cuvier, the logic structuring organisms was understood, one could not avoid, in order to explain their functioning, looking towards the outside world: does the organism not illustrate adaptation to an environment which it needs, but which threatens it by multiple pressures? So one learns to seek out the relationships between the being and the environment: that is where geography can find its place. Humboldt was the first to feel this. Did he not write, on 5 June 1799, in the boat which was ready to take him from Coruna towards America: 'I shall try to find out how the forces of nature interact with one another and how the geographical environment influences plant and animal life. In short, I ought to discover what makes the unity of nature' (quoted by Botting, 1973, 65). So from the outset geography, which also discovers problems of organization and structure, is seen more as an environmental than a human science, whilst the seventeenth- and eightenth-century traditions and the works of mercantilists and statisticians could have made it swing to the side of economics.

(c) Michel Foucault notes, midway through the modern period of the *episteme*, a change in perspective (Foucault, 1966): the assumptions controlling intellectual progress as a whole are not called into question, but they are no longer applied in the same way. From then on the insistence is more on the notion of the norm rather than on function, on the rule rather than on conflict, on the linguistic system rather than on meaning. One goes from a world where the historicist preoccupation is dominant to a world which is more interested in the articulations of reality, in control mechanisms and in structural assemblages. In such a climate, curiosity about space once more predominates: such curiosity is no longer limited to the question of the relations between the being and the environment; it can take the broader approach of the spatial constraints affecting man, and give the social sciences new coherence.

The chronology of this change is indicated in a very vague way by Foucault: he sees the first signs of it in some late nineteenth-century philosophers, Nietzsche, for instance; however, as regards modern times, he is unfortunately much better informed about philosophical development than about the evolution of the social sciences. Moreover, he does not hide the fact that he has small regard for them: they cannot all achieve the status of rational and objective knowledge. He therefore compares the trilogy with which he chose to follow the evolution since the seventeenth century (life sciences, linguistics and economics) with other disciplines: sociology, history and ethnology, for example. He considers the first three to be true sciences because they have been able to create a perfectly definable object, which is by nature objective; life, work, language in its material manifestations. The others are closely related to these but in a space where man is no longer absent – in a field where the problem of explanation and comprehension arises from the outset. In this field, for Foucault, it is not possible to design sciences with firm structures, but it is essential to see that structured knowledge is involved.

The analysis which Foucault proposes undoubtedly contains an element of the arbitrary. The comparison between economics and social sciences will not convince those who practise them and who see that it is more and more difficult to isolate and to understand them without constantly passing from one field to the other. Foucault remains the prisoner of a certain Marxist viewpoint and has not succeeded in breaking free from the presuppositions of the analysis of the value of work. But that does not at all detract from the scope of his proposal. By placing the originality of humanities on a different level from that of the traditional explanation/comprehension comparison, he enables a problem to be overcome which has long been an obstacle to the epistemological strengthening of these disciplines: he invites their analysis as forms of knowledge, i.e. as disciplines constructed on practices and on perfectly defined and coherent techniques, even if they do not have the privilege of the support of a perfect base for the epistemologist. And thus he is launched on his great quest for the hidden relationships between power, social order and the social sciences. *Surveiller et punir* (Foucault, 1975) is his first major contribution in this direction. Even if it is largely outside the sphere which could interest the geographer, everything he says on the techniques of partitioning space, observation or on Bentham's *panopticon* economics leads us in the direction of the unjustly forgotten reflections of the statisticians, the legal writers and political economists by which Western society has gradually obtained the instruments of action necessary to it and at the same time gave rise to the correlative scientific disciplines.

So Foucault's analysis invites one to go beyond the history of geographical thought, far beyond academic spheres, e.g. those of economists, historians or ethnologists, who have been mentioned previously: it leads one to draw up a list of the spatial practices of European societies, to determine the way in which statesmen, soldiers, financiers and tradesmen think about space, use it and transform it: such is the stuff of geography. From the outset it has been intimately bound up with the action it has guided. The problem with academic geography is often to explain why concepts and theories have lagged behind practice, rather than the reverse.

With this bias, the history of geographical thought converges on the other area of modern epistemological concern – the one opened up by Gunnar Olsson (1975) when he investigates the logic of geographical practices and tries to understand how axiological reasoning differs from explanatory reasoning in which up to now critical thought has been too exclusively interested.

Note

1 After drafting this article I discovered the study by Robert D. Sack (1976) on magic and space. This is the best illustration of Foucault's themes of epistemology with regard to geography in the framework of the *episteme* of the Renaissance.

References

Althusser, L. 1969: Avertissement au lecteur. In Marx, Karl, *Le Capital* (Paris) 7–26.
Althusser, L. and Balibar, E. 1968: *Lire le capital* (Paris), 2 vols., 184, 228.
Bachelard, G. 1934: *Le nouvel esprit scientifique* (Paris).
Bachelard, G. 1938: *La formation de l'esprit scientifique* (Paris).
Bartels, D. 1968: *Zur wissenschaftstheoretischen Grundlegung einer Geographie des Menschen* (Wiesbaden).
Botting, D. 1973: *Humboldt and the Cosmos* (London).
Buttimer, A. 1971: *Society and milieu in the French geographic tradition* (Chicago).
Claval, P. 1964: *Essai sur l'évolution de la géographie humaine* (Paris).
Downes, A. 1971: The bibliographic dinosaurs of Georgian geography, 1714–1830. *Geographical Journal* 137, 379–87
Foucault, M. 1966: *Les mots et les choses* (Paris).
Foucault, M. 1969: *L'archéologie du savoir* (Paris).
Foucault, M. 1975: *Surveiller et punir* (Paris).
Hard, G. 1969: Die Diffusion der 'Idee der Landschaft'. *Erdkunde* 23, 249–64.
Hard, G. 1970: Die 'Landschaft' der Sprache und die 'Landschaft' der Geographen. *Colloquium Geographicum* 11, 1–278.
Hartshorne, R. 1939: The nature of geography. *Annals of the Association of American Geographers* 29, 171–658.
Hartshorne, R. 1959: *Perspective on the nature of geography* (Chicago).
Harvey, D. 1969: *Explanation in geography:* (London).
Kuhn, T. S. 1962: *The structure of scientific revolutions* (Chicago).
Levy, J. 1975: Editorial. *Espaces Temps*, no. 1.
Popper, K. 1945a: *The open society and its enemies.* (London).
Popper, K. 1945b: *The poverty of historicism* (London).
Olsson, G. 1975: *Birds in egg.* Ann Arbor, Department of Geography, University of Michigan.
Reynaud, A. 1971: *Epistémologie de la géomorphologie.* (Paris).
Reynaud, 1974: *La géographie entre le mythe et la science. Essai d'épistémologie.* (Reims).
Robinson, B. S., 1973: Elizabethan Society and its named places. *Geographical Review* 69, 322–33.
Sack, R. D. 1976: Magic and space. *Annals of the Association of American Geographers* 66, 309–22.
Schmithusen, J. 1970: *Geschichte der geografischen Wissenschaft von den ersten Anfängen bis zum Ende des 18 Jahrhunderts* (Mannheim).
Schmithusen, J. 1976: *Allgemeine Geosynergetik* (Berlin).

Index

Abbagnano, N., 51
Abel, T., 110, 127n
Abercrombie, P., 194, 200
Aberdare, Lord, 57
Abrams, P., 188, 204n
academic discipline:
  geography's
  development into, 17; see
  also university
academic freedom, 'pure'
  idea of, 169
academy, and
  institutionalized
  structure of science, 28
African Association 54
Agassi, J., 2, 3, 8
age of enlightenment, 29
agriculture, Reclus and,
  160
Algeria, 163
Allen, W., 184n
Almagìa, R., 44
Althusser, L., 229, 230, 236
American Geographers,
  Association of, 72, 75
American Geographical
  Society, 60
anarchism: Geddes and,
  187, 196; Kropotkin and,
  134ff; Reclus and, 155–63
Andersson, G., 18, 30
Anordung (experienced
  relationships), 105
Anscombe, G. E. M., 110
anthropology, 193; text
  interpretation and, 122
Anuchin, V. A., 9
Appleton, J., 122
Arago, F., 42

archaeology, 40
Archéologie du savoir, L'
  (Foucault), 182n, 231,
  233
areal differentiation,
  geography as, 99
Aron, R., 177, 178, 179,
  184n
astronomy, 58
atomic physics,
  geographical reasoning
  and, 4
Ausdruck (expression), 107,
  116
autobiographical
  perspective, 86ff
Avakumović, I., 158
Avrich, Paùl, 163
Bachelard, G., 11, 122, 229,
  236
Bacon, Francis, 43
Baker, J. N. L., 2, 43
Baker, P., 213
Bakunin, Michael, 155, 156
Balbi, Adriano, 41, 44
Baldwin, R., 143, 147
Banse, E., 31, 100, 110, 113,
  128n; and
  Landschaftskunde school,
  114, 115
Barbour, B., 77
Barnbrock, Jörn, 177, 181,
  182
Barnes, H. E., 88
Barrett, C., 214
Barron, F. X., 95
Barrows, Harlan, 99
Bartels, Dietrich, 3, 25, 30,
  230

Bates, H. W., 144
Beaver, S. R., 202
Beck, H., 21, 40
Beck, L. W., 129n
behavioural geography, 73
Behr, K. M., 45
Ben-David, J., 12, 62
Berdoulay, V., 3, 5, 12, 13,
  14
Berg, L. S., 31
Berger, H., 21
Berger, P., 129n
Bergson, Henri, 192
Berkeley (University of
  California), 211, 212,
  214, 224n; influence of
  Berkeley school, 71–2, 73
Berlin, 172; University of,
  39, 155, 173
Bern, 60, 62
Bernal, J. D., 11
Bernstein, R. J., 167
Berry, B. J. L, 11, 30, 71, 72,
  75, 81
Bertacchi, C., 44
Besinnliches Nachdenken,
  84, 90, 91
Beveridge, W. I. B., 77
Bibby, C., 77
Bildung, universities and,
  169
Billinge, Mark, 6, 182
biology, 236; Geddes's
  work in, 203, 204n
Biometrica, 190
Bismarck, Otto von, 169
Blaug, M., 74
Boardman, P., 187, 200
Bobek, H., 113, 114

Böhme, G., 18, 30
Bollnow, O., 100, 101
Booth, Charles, 188, 191, 192, 197, 205n
Bortoft, H., 119, 129n
Boston, 123, 144
botany, 40; and medicine, 29
Botting, D., 237
Bowden, M., 122
Branford, Victor, 189, 205n
Brazil, 5
Breitbart, Myrna Margulies, 5, 149
Brigham, A., 210
British Geographers, Institute of, 76
Broc, Numa, 41, 42, 55, 59, 165
Brod, Max, 173
Brodbeck, M., 208
Brooke, M. Z., 193
Brown, S. G., 219
Brunhes, Jean, 62–3
Brussels, 161; Free University, 156; New University, 156
Bruyn, S., 129n
Buber, M., 101, 110
Bukharin, N. I., 11
Bunge, W., 76, 100
Buranelli, V., 211, 212
Burckhardt, J., 10
bureaucratization: Alfred Weber's essay, 183n; Max Weber's concern with, 176
Burton, I., 74, 75, 100
Butterfield, H., 11
Buttimer, Anne, 5, 23, 25, 34, 86, 100, 115, 182, 208, 209, 230; Dilthey and, 116, 122, 129n
Büttner, M., 11, 28
California, 63; Royce and, 210, 211, 212, 213, 220, 221, 222, 224n
Camden, William, 234
Campos, Tóres, 53
Capel, Horacio, 5, 39, 64, 67
Capital (Marx), 229
capitalism: community action and, 196; and geographic thought, 11; Kropotkin's rejection of,

138, 139, 140; and location theory, 166, 176, 177; political economy and, 187
Carew, Richard, 234
Carlyle, Thomas, 101
Carnegie Trust, 200, 219
cartographic tradition, 21, 47, 235
Casati law (1859), 44
Cayley, G., 75
Cherry, G. E., 200, 201
Chicago, 86, 191, 199, 204n, 205n, 209, 213
Chickering, R., 177
Chorley, R., 11, 26, 100, 223n; paradigm idea, 71, 72, 73, 81
chorographical tradition, 21
chorological tradition, 99
Christaller, Walter, 75, 76, 174
chronology: emphasis on, 2; and history, 39, 44
Cipolla, C., 48, 50
'circle of affinity', 14
Cities in evolution (Geddes), 190, 197, 204n, 205n
civics, Geddes and, 187, 190, 191, 192, 193, 201
Civics as applied sociology (Geddes), 190
Clairvaux, 144
Clapham, J. R., 183n
Clark, T. N., 12
Claval, Paul, 4, 9, 10, 14, 41, 182, 182n
Clay, G., 121, 122
Clandenning, J. J., 212, 213, 214
climatology, 47
Coale, George, 211
cognition, 216
cognitive interests, Habermas's notion of, 178
'cognitive map', of environment, 25
Cole, T., 183n
Coles, R., 122
'collective consciousness', Royce's concept of, 212, 220, 223
Collège des Ecossais, 204
Collingwood, R. G., 101,

116, 124n
Collins, R., 62
colonialism: and geographical societies, 58, 59; impact on geographic thought, 11, 30, 53, 60; Reclus's opposition to, 163; see also imperialism
commerce, development of, 58
commune: Kropotkin's concept of, 147; see also community
communication, Habermas's views on, 168
Communist Party, Weber and, 180
community: and environment, 146; Geddes's concept of, 195; Royce's concept of, 208–15, 217–19, 222, 223; see also provincialism
Comte, Auguste, 101, 189, 191
consciousness, Kant's concept of, 102
context: contextual approach, 8–14; development of geographical thought, 3, 8; paradigm and, 88; see also milieu
continental drift, paradigm idea and, 72
contingency, and geographical thought, 3
'conventionalism', 9
conurbation, Geddes's concept of, 194
Copernicus, 71
Cornish, V., 199
Cortambert, Eugène, 42
Cosmos (Humboldt), 40, 42, 43
Costello, H., 213
Cotton, J. H., 217
Craik, K., 122
Critique of pure reason (Kant), 101
Croce, B., 101, 110, 124n
Crone, G. R., 41, 43, 47, 55, 58
cultural sciences,

'individuating'
methodology of, 104
cumulation, emphasis on,
2
Dardel, E., 122
Darwin, Charles, 2, 3, 45,
137, 189; organization of
knowledge, 4; slavery in
Brazil, 5
Dasein, 84
data matrix, 71
Davis, William Morris, 72,
76, 144, 154; and Royce,
213, 217, 223n
decentralism: Geddes and,
195; Kropotkin's theory
of, 135, 139, 140–42, 146–
9, 150, 151; Royce and,
222
de Dainville, F., 10
Defries, A., 196
demography, 189
Demolins, F., 193
denudation, 72
dessication, Kropotkin's
theory of, 134
determinism, 11
'developing countries',
state of geography in, 12
development,
reconstruction of, 19
Dewey, J., 101
Dickinson, R. E., 2, 5, 43,
47, 60, 75, 99, 114, 128n,
199, 202
diffusion theory, 77
Dilthey, Wilhelm, 5, 100,
126n, 127n, 128n, 129n,
172, 179; concept of
Verstehen, 106–8, 129n;
'critique of historical
reason', 101–2; *Erlebnis*,
104–6, 125n;
Geisteswissenschaften,
102–4; hermeneutics,
108–10; Hettner, 111–12;
human geography,
110ff; Kraft, 112–13;
Landschaftskunde school,
113–15; modern
geography, 115–17;
notion of 'text', 117–20,
123, 124; and Ritter, 110–
11
disciplinary divisions,
crisis of existing, 37
discourse: concepts

through which
conducted, 166; events
through which effective,
167; process through
which produced, 166
Dockès, P., 165
Dokuchayev, V., 31, 45
Downes, A., 235
Downs, R., 100
Drache, H. M., 160
Droysen, J. G., 125n
Dubois, J. J., 60
Dubois, Marcel, 52, 53
Duhem, Pierre, 9
Dunbar, G. S., 5, 156, 160
Duncan, S. S., 76
Dunfermline, 200, 201
Durant, Will, 158
Dussieux, L., 42
dynamics, Geddes and,
191, 198–9
ecology, 32, 51, 99, 100;
Kropotkin's theory of,
135, 139, 150; Park's
human, 209
economic growth, 32
economics, 52, 238;
geography and, 45;
paradigm idea in, 72, 73
economy of scale,
limitations on, 138
Edge, D. O., 12
Edie, James, 129n
Edinburgh, 191, 195, 201,
204; Royal Society, 197;
Social Union, 196;
University of, 144
education: advances in
popular, 59;
cameralistic, 169;
diffusion in 19thC,
48–51; Geddes and, 190,
202; geography and, 3,
38, 47; Kropotkin on,
139, 145–6, 150; reform,
53, 54, 76; *see also*
academy, university
Edwards, Jonathan, 212
Edwards, K. C., 202
egoism, Royce's concept
of, 217
Einstein, Albert, 71, 74
elitism, Weber and, 181
Emerson, Ralph Waldo,
212
empathy, *cf. Verstehen*, 106
empirical science, 25

empiricism, 101
Entrikin, J., 5, 129n, 182
environment, 28, 31, 32,
237; cognized, 23, 25, 31,
33; community and, 146;
Geddes's view of, 201;
man's consciousness of,
23; perceived, 23, 31; real
physical, 23; Royce's
view of, 220, 221, 223
environmental
determinism, 72
environmental planning,
90, 198
environmental
possibilism, Royce's
philosophy of, 221
environmental sciences, 32
environmentalism, 31
epistemology: Foucault's
categories, 231–4;
progress of geography,
227–31, 234–8
Erdkunde (general earth
science), 100, 111, 113
Erdwissenschaft, 128n
Erfahrung (experience in
general), 101, 105, 126n
Erfurt, 172
Erlebnis, Dilthey's concept
of, 101, 104–9, 111, 113,
115, 116, 125n, 126n,
127n; *in situ*, 121, 124
Erlebnisausdruck (life
expressions), 109, 120
ethics, neglected issue, 13
Ethics (Kropotkin), 137, 147
ethnography, 58
ethnology, 43
Eucken, R., 171, 174
eugenics, 189, 190
Eugenics Education
School, 190
Eugenics Laboratory, 190
Eugenics Review, 190
Evans, Estyn, 202
evolution: Geddes's
theories, 193, 203;
mutual aid and, 137, 138
evolutionism, 64
exceptionalism, 228
'exceptionalist' tradition, 2
existentialism, 100, 101,
110; Royce and, 223
experience: and
knowledge, 23, 83, 84;
and perception, 105

explanation, methodology, 4
Explanation in geography (Harvey), 71
exploration: geography and, 30; narrative of change, 2; tradition, 29, 47, 58
expression, as intermediary, 107, 109
'externalist' bias, 12
'extraordinary science', geography and, 82, 85, 87–9
family, Le Play's concept of, 191
Faris, R. E. L., 204n
fascism, Weber and, 179
Faure, Elie, 163
Faure, Sébastien, 157
Fawcett, C. B., 199, 202
Febvre, L., 5
federation, Kropotkin on, 141, 149
Ferkiss, V. C., 87, 88
Feud of Oakcreek, The (Royce), 212
Feyerabend, P. 87
Finland, Kropotkin in, 134, 135, 144, 147
Fischer, E., 99, 114, 128n
Fleure, H. J., 192, 201, 202
Forster, J. R., 47, 111
Foucault, Michel, 4, 11, 12, 165, 182n, 230, 231–8, 239n
France, 31, 59, 162, 192, 229, 230, 235; conventionalism in, 9; education, 48, 53; ethics, 13; geographical societies, 55, 59; institutionalization of geography in, 41–2; Kropotkin imprisoned in, 144; *pays*, 87; rationalism, 168; Reclus in, 154, 155, 161; spatial dimension, 165
Franco-Prussian War, 3
Frankel, H., 72
Frankfurt, 39
Frankfurt school, 86
Franklin, F., 212, 224n
Free Socialism, Weber and, 180
Freeman, T. W., 5, 6, 42,

43, 54, 55, 58, 59, 182
French Revolution, 41, 52, 167
Freud, Sigmund, 2
Friedrichs, C. J., 183n
Frisby, D., 172
Fröbel, Friedrich, 51
Gadamer, H. G., 117, 118, 119, 120, 121, 123, 129n, 130n
Gaidoz, H., 159
Galois, B., 140
Galton, Francis, 189, 190
garden city movement, 195
Gardiner, P., 125n
Garnett, A., 202
Gauss, F., 46
Geddes, Patrick, 144, 148, 161, 162, 163, 186, 189, 204; anarchism, 187; concept of community, 195–6; doctrines, 190–94; dynamics, 198–9; holism, 198; influence on geography, 201–3; legacy, 200; planning, 200–201; social structure, 196–8; unrealized contribution of, 194–5; urban focus, 199
Geisteswissenschaften (systematic human studies), 100, 169; Dilthey's concept, 102–4, 111, 112, 114, 115, 117, 125n
Gellner, E., 87
Gendlin, E., 129n
general systems theory, 33
genres de vie, in geographic perspective, 94, 95, 96
geodesy, 47
geographers, community of, 38; growth of, 59–62; strategies of, 62–5
Geographia generalis (Varenius), 39
Geographical Association, 202
geographical consciousness, 22
geographical enquiry, structure and process of, 22–6
geographical society, evolution of, 47
Geographical Society of

Madrid, 59
geographical thought: context of change in, 75; history of, 82ff
'géographie humaine' tradition, 99, 100, 122
geography-of-nature tradition, 29
geologist, and community of geographers, 62
geology, 82; development of, 46; Le Conte's courses, 224n; Museum of Applied Geology, 43; and physical geography, 43
geomagnetism, 40, 46
geomorphology, 72, 73, 82; coherence of, 230
Geomorphology (Von Engeln), 73
geophysics, 40, 46
Georg, Stefan, 176
Gerasimov, I., 45, 46
German Ideal of Freedom, The (Krieger), 183n
Germany, 40, 43, 159, 230; Davisian geomorphology, 73; education, 48, 59; geographical societies, 55; industrialization, 170–71; institutionalization of geography in, 21, 29, 30, 40–41; 'Landschaft', 31; location theory, 165; Reclus in, 155; romanticism, 169; Royce in, 211, 212; urban growth, 175, 183n; Weber in, 166, 172–3, 176, 183n
Gesellschaft für erdkunde zu Berlin, 54
Gesellschaftsprozess (societal process), Weber's notion of, 177, 178
gestalt, in scholar's life, 90
Gestaltende Geographie, 114–15
Giblin, Beatrice, 162–3
Giddens, A., 168
Gilbert, E. W., 202
Gilman, Daniel Coit, 212, 213, 224n
glaciation, Kropotkin's

theory, 134, 147
Glacken, Clarence, 6, 10, 99
Gladwin, T., 122
Glass, R., 199
Goethe, Johann von, 40, 101, 126n
Goodman, Paul, 148, 150, 151
Goodman, Percival, 148
Gould, P., 75, 229
Granö, J. G., 28
Granö, Olavi, 5, 10, 25, 26, 30, 83, 85, 86, 96
Grave, Jean, 157
gravity, 47
Great Britain, 159, 186, 193; colonialism, 60; Davisian geomorphology, 73; education, 48, 50; geographical societies, 55, 59; industrialization, 169–70; institutionalization of geography in, 42–3; Kropotkin lectures in, 144; physiography, 51
Green, M., 174, 176, 177, 183n
Gregory, D. J., 4, 5, 6, 73, 166, 167, 168
Grigg, D., 71
Gottmann, Jean, 194
Guelke, L., 115, 124n
Guerin, D., 147
Gusdorf, G., 11, 12
Gustavsson, S., 18
Guyot, Arnold, 214, 224n
Habermas, J., 86, 168, 178
Hägerstrand, Torsten, 19, 34, 76, 82, 87, 129n
Haggett, P., 9, 11, 26, 71, 72, 73, 75
Hahn, R., 8, 12
Haldane, J. B. S., 11
Hall, D. H., 46, 47
Hallam, A., 72
Halle, University of, 169
Halliday, R. J., 189
Hard, Gerhard, 4, 21, 25, 30, 31, 230
Harris, Cole, 115, 116; historical synthesis, 124n
Hartshorne, R., 2, 4, 9, 41, 76, 99, 100, 230; and Dilthey, 115

Harvard, Royce and, 210, 211, 212, 213
Harvey, D., 71, 76, 100, 209, 228
Hausmann, Georges, Baron, 195
Hawthorn, G., 169, 172, 180
Heelan, D., 118–19
Hegel, Georg, 100, 104, 111, 169, 177, 219
Heidegger, M., 18, 83
Heidelburg, 174, 176, 179, 180, 182n, 183n
Heimatkunde, 51
Heisenberg, 4, 74
Heller, O., 84
Hempel, C., 85, 86
Herbert, D. T., 72
Herbertson, A. J., 31, 162, 201, 202
heredity, eugenics and, 189
hermeneutics, 101; Dilthey and, 108–10, 117, 118, 126n
Hermeneutik (Schleiermacher), 128n
Hermeneutik und Kritik (Schleiermacher), 128n
Herodotus, 42
Hessen, B. M., 11
Hettner, Alfred, 2, 29, 31, 41, 73, 99, 100, 110; and Dilthey, 111–12, 113, 115, 128n
Hindess, B., 182n
Hirshman, C. F., 184n
historical development, division into periods, 20
historical method, sharpening of, 1
historiography: foundations of, 19–21; of geography, 17; origin of, 21–2
history: development of human consciousness, 8; Geddes's view of, 192; geography as ancillary to, 28, 29, 39, 41, 44, 45, 48, 51; of ideas, 3; normative quality of, 2
History of Geographical Thought, Commission on the, 1, 6
Hobhouse, L. T., 189, 190

Hodges, H. A., 104, 105, 110, 125n
Hoffman, R., 136
Holborn, H., 102
holism: Geddes and, 198, 202; Royce and, 209, 223
Holloway, W., 147
Homer, 42
Homme et la Terre, L' (Reclus), 160, 161, 163
Hooson, D., 10
Howard, Ebenezer, 148, 195
Hughes, H., 213
Hugo, Victor, 144
Hull, D. L., 76
human geography, 31; Dilthey's study of, 110ff; as text-interpretation, 120ff
human studies, Dilthey's concept of, 102–4, 125n
humanist perspective, Royce and, 209
'humanistic' movement, 23
Humboldt, Alexander von, 2, 5, 9, 29, 39–40, 41, 42, 43, 59, 154, 169, 237; concept of natural unities, 47; 'father of geography', 66; orographic structure of Asia, 147; theory of the earth, 46
Hume, David, 101, 102
Hurst, M. E. E., 25
Husserl, Edmund, 101, 108, 109, 126n, 127n, 128n, 210; and Weber, 179
Huxley, Thomas, 51, 77, 187
idea: environmental appropriateness, 87; evolution of, 8, 10, 84; history of, 88–9, 96
idealism, 100, 101, 124n, 178; Royce's concept of, 211, 212–14, 223; transcendental, 209; Weber and, 180
ideological setting, importance of, 83
ideology, 4
imperialism, 53, 55; see also colonialism

India, 192, 195
individualism, Royce's
  concept of, 217
industrial revolution,
  education and, 50
industrial transformation,
  intellectual traditions
  and, 167–72
industrialization, 30, 46;
  and family, 191;
  Germany, 170–71;
  Kropotkin's views, 138
'Initial credibility',
  characteristic to text, 122
Innewerden (inner
  perception), 105;
  Dilthey's concept of,
  126n
innovation, 13, 88;
  institutionalization of,
  12; nature of, 77, 78
institutionalization;
  development of
  geography in context of,
  26–34, 46; of innovation,
  12; of 'new geography',
  47ff; of sciences, 37–8;
  stages of, 13
intellectual change,
  process of, 5
intellectual traditions, and
  industrial
  transformation, 167–72.
intellectual trend,
  importance of general,
  83
intentionality, Husserl's
  theory of, 109, 126n
International
  Geographical
  Conference: Bern (1891),
  60; London (1895), 53
International
  Workingmen's
  Association, 155
interpretation, and process
  of knowing, 216
*Introduction to the Human
  Studies* (Dilthey), 124n
*Introduction to the science of
  society* (Park and
  Burges), 204n
introspection, *cf. Verstehen*,
  106
intuition, and process of
  knowing, 216
investigation, formal
  structure of, 71

Isard, W., 174, 183n
Italian Geographical
  Society, 59
Italy, 48, 67;
  institutionalization of
  geography in, 43–4
Jackson, J. B., 121, 122
Jaffe, Edgar, 174
James, P. E., 3, 5, 29, 75,
  99, 100, 111, 114
James, William, 101, 110,
  210, 211, 212, 213
Janik, A., 88
Jaspers, K., 101, 171
Jeffreys, Harold, 72
Johns Hopkins University,
  212
Johnston, R. J., 71, 72, 73,
  182n
Jones, E., 208
Jugend im Nebel (Brod), 173
Kaila, E., 33
Kant, Immanuel, 2, 9, 39,
  101, 102, 126n, 169, 173
Kausalzusammenhang
  (causal system), 128n
Keltie, John Scott, 59, 60;
  and Kropotkin, 143, 144,
  150
Kende, O., 128n
Kimmerle, H., 128n
King, L. J., 75
Kirk, W., 4
Kitchen, M., 170
Kitchen, P., 191
Kitts, D. B., 72
knowledge: action
  extrapolated from, 23;
  attitude to, 95;
  development of, 87; and
  experience, 83, 84;
  integration of, 91;
  language and, 231, 232;
  Royce's concept of, 216,
  217
Kockelmans, J. J., 18, 127n,
  210
Koestler, A., 87, 88
Koyré, A., 11, 67
Kraft, Viktor, 100; Dilthey
  and, 112–13, 128n
Kretshmer, K., 41
Krieger, L., 183n
Krise der Wissenschaft, 172
Kropotkin, Peter, 5, 134–6,
  150–52, 196;
  decentralism, 140–42,
  146–9; as geographer,

143–5; moral principle in
  nature, 136–8; on
  purpose of geography,
  145–6
  and Reclus, 156, 157, 158,
  161, 163; social
  revolution, 142–3;
  struggle for human
  expression, 138–40
Kuhn, T., 11, 12, 18, 25, 67,
  165; extraordinary
  science, 87; milieu, 90;
  normal science, 85;
  notion of paradigm,
  70–73, 74, 76, 77, 78, 81,
  84; structure of scientific
  revolutions, 229
Kuklick, B., 210, 211, 215
Kulturbewegung (culture
  movement), 178
*Kulturgeschichte als
  Kultursoziologie* (Weber),
  177
Kulturlandschaft (cultural
  landscape), 114
Kultursoziologie (cultural
  sociology), historical
  specificity, 166
Kulturstaat, doctrine of, 169
Kulturwissenschaften
  (cultural sciences), 100,
  112, 126n
La Blache, V. de, 198
Lacoste, Y., 11
Lakatos, I., 11, 74, 78, 81,
  87
land-centred approach, 25
Länder, 47
Landerkunde, 100, 111,
  112, 113
Landes, D., 171
landform: evolution of
  Davisian, 82; paradigm
  for historical analysis of,
  72
land redistribution,
  Kropotkin's theory of,
  148
land registry, Russia, 45
Landschaft, 4, 31
Landschaftskunde, 110,
  113–15
land use, Kropotkin's
  theory of, 139
language, and
  epistemology, 231, 232,
  233, 236
Lauer, Q., 128n

Lausanne school, 182n
Lautensach, H., 110, 113
Leach, Edmund, 234
Le Conte, Joseph, 214, 224n
Lecourt, D., 182n
Lécuyer, B-P., 11, 12
Lee, J. J., 183n
Lee, Roger, 182
Lefèbvre, H., 90
Lehmann, O., 4
Leighly, J., 72, 73
Lenntorp, B., 19
Le Play, Frédéric, 99, 191, 193, 196, 203, 205n
Le Play Society, 202
Levasseur, Emile, 14, 53
Levi-Strauss, Claude, 234
Levy, J., 228
Lewis, D., 122
Lewis, J., 167
Ley, D., 122, 209
Liebow, E., 122
'life expression', 125n
Life of Schleiermacher, The (Dilthey), 124n
life-philosophers (Lebensphilosophie), 100, 101, 124n, 172–4
Linné, Carl, 29, 232
lived experience, Dilthey's components, 104
location theory, 165–6, 171, 174, 176–8, 181, 182
Locational analysis in human geography (Haggett), 72, 75
Locke, John, 101, 102
logic, 227, 228
Lomonosov, M. V., 45
London, 54, 62, 144, 188, 191, 194; University College, 42; School of Economics, 189
Lotze, Herman, 211, 212
Louch, A., 87
Lovejoy, Arthur, 10
Lowenthal, D., 4, 6, 100, 115, 122, 129n
Lucke, F., 128n
Lüdde, J. G., 21
Lukermann, F., 4, 11, 115
Luoma, M., 184n
Lyell, C., 45, 46
Lynch, K., 121
Mach, Ernst, 4

MacIver, B. M., 84, 95
Mackin, J. H., 76
Mackinder, H. J., 31, 75, 76, 189, 194, 201
Mackintosh, Charles Rennie, 194
MacLeod, R., 18
Maconochie, Captain, 42
Magnetic Union, 46
Mairet, P., 196
Maitron, Jean, 157, 158
Makkreel, R., 103, 104, 108, 109, 110, 111, 125n, 126n, 127n, 129n
Malatesta, Errico, 150, 157
Malato, Charles, 157
Malte-Brun, Conrad, 41, 42
man-centred approach, 25
Manchester, 188, 204
Mannheim, K., 101, 110
Marcel, Gabriel, 223
Marcuse, H., 86, 168
Marr, J. S., 73
Marr, T. R., 204
Marx, Karl, 2, 100, 101, 229
Marxism, 100, 157, 187; epistemology and, 228, 236; scientific ideas and, 11
Marxism and epistemology (Lecourt), 182n
Maslow, A., 110
Mason, E. B., 97
Massey, D., 166, 181
Masterman, M., 73, 82
materialist concept, of geography, 9
Mattson, Kirk, 67
Matveyeva, T. P., 45
May, J. A., 22, 129n
McCarthy, T., 167
McDermott, J. J., 211, 212
McMurray, J., 88, 91, 96
McRae, D., 179
Mead, G. H., 211
Mellor, H. E., 192, 201
Melville, B., 167
Merleau-Ponty, Maurice, 118
Merton, R. K., 12, 18
Mesnil, Jacques, 157
meteorology, 40, 47; creation of stations, 58
Methodenstreit, 172, 175
methdology, and

geography, 2, 8, 10
Mexico, 63
Meyer, D., 72
Meynier, André, 9, 42, 60, 162
Michelet, Jules, 158
Michelson, W., 122
Milan, 44
milieu, geographical thought and, 81, 83, 88, 90–91, 96
Mill, H. R., 73
Mill, J. S., 125n
mineral resources, exploitation of, 46
mineralogy, 47
Minguet, C., 40
Mitscherlich, A., 180
Mitzman, A., 176
model-building, 100
'models of intelligibility', 12
Moffatt, I., 72
Mogey, J., 193
Moles, A., 122
monism, 23, 31
Moore, R. L., 160
Morgan, Lloyd, 192
Muirhead, J., 220
Mulkay, M. J., 12, 18
Müller-Wille, C. F., 75
Mumford, Lewis, 148, 158, 162
Münsterberg, Hugo, 210
Murchison, R. I., 40, 43; Royal Geographical Society, 55
Murphy, G., 88
Musgrave, A., 11, 81, 87
mutual aid, Kropotkin's theory of, 135–40, 147
Mutual Aid (Kropotkin), 147
Myrdal, G., 86
Nacherleben, 107; Dilthey's concept of, 129n
Nagel, E., 110, 127n
Napoleon III, 155
Nardy, J-P., 9, 14
National Association for the Promotion of Social Science, 189
National Geographic Society, 221
nationalism, geography and, 51–4

natural sciences, *cf.* human studies, 103, 125n
nature: contact with, 50; Darwinian view of, 137; geography of, 28; unity of, 40
Nature, 144
Nature of geography, The (Hartshorne), 230
Naturwissenschaften, 122, 125n
navigation, maritime, development of 46, 47
Nazi seizure of power, The (Allen), 184n
Nazism, Weber and, 179
Needham, J., 11
Netherlands, 50
Nettlau, Max, 158
Neuchâtel Geographical Society, 162
Neumann, S., 177, 178
Neuwied, 155
'new geography', 229; introduction of, 75
Newman, O., 122
Newton, Isaac, 71, 74
New Towns, 87
Nicod, J., 33
Nietzsche, Friedrich, 100, 101, 137, 237
Nineteenth Century, The, 144
noetic-neomatic structure, in text, 122
'normal science', geography and, 81, 85–7
Nossiter, B. D., 186
Nouvelle géographie universelle (Reclus), 156, 158, 161
Oberschall, A., 12
objectivity, 'psychic' *cf.* 'natural', 109
observation, education centred upon, 51
observatories, creation of, 47
Olsson, Gunnar, 76, 238
Ordnance Survey, 58
Origin of Species (Darwin), 137
Ormsby, H., 202
orography, Kropotkin's theory of, 134, 144, 147
Oxford, University of, 42, 59, 144, 145, 199, 202

Paassen, C. van, 22, 129n
Pahl, R. E., 208
Pallas, M., 47
Palmer, G. H., 210
Palmer, R., 100, 105, 109, 126n
paradigm: applicability of, 72–4; change, 74–7; geographical, 71–2; Kuhn and, 229; notion of, 11, 25–6, 70–71, 78, 81
Paris, 42, 156
Paris Commune, 155
Park, R. E., 191, 204n, 209, 210, 213
Parsons, J. J., 72
Passarge, S., 31, 110, 113
patria, *see* nationalism
Pattern and Meaning in History (Rickman), 127n
Pattison, W. D., 9
pays, regionalization scheme based on, 87
Pearson, Karl, 190
pedology, genetic, 45
Peirce, Charles, 212, 215, 216, 220
Penck, Albrecht, 73, 100
Penck, Walther, 73
'peneplain', 81
perception, 229; Dilthey's concept of, 103, 105; and process of knowing, 216
perceptual-cognitive studies, 31
perceptual geography, 32
Perry, R. B., 211
Perspective on the nature of geography (Hartshorne), 230
Peschel, O., 21
Pestalozzi, Enrico, 28, 50, 101
phenomenological basis, of geographical science, 22
phenomenology, 101, 108, 109, 110, 179; Royce and, 210; of science, 18
Philosophy and methodology in the social sciences (Hindess), 182n
philosophy, natural, 29
physical geography, 31; paradigm change, 78
Physical Geography (Somerville), 43

physics, 33, 40, 46
Physiographie (Cortambert), 42
physiography, 51
Piaget, J., 11
Pick, George, 174
Pinchemel, Philippe, 6
Pivcević, E., 179
Planck, Max, 4, 74, 76
'Planck principle', 76
planning, town, Geddes and, 188, 189, 195, 200–201, 203
plate tectonics, 73; paradigm idea and, 72
Plewe, E., 3
Poincaré, Henri, 4, 9
Polanyi, M., 86
political economy, 45; ameliorism and, 188; epistemology and, 228, 236; Kropotkin's definition, 141; location theory and, 166; nineteenth century, 187; positivism in, 186; spatial dimension, 165
politics, geography as ancillary to, 28, 29
Pomeroy, E., 211, 221
Popper, Karl, 67, 74, 78, 86, 228
Popular Science Monthly, 223n
population, Reclus's views on, 161
Porteous, J., 122
positivism, 31, 64, 100, 165, 174, 204n, 209, 227, 231; dominance in political economy, 186, 188; influence in pedagogy, 51
poverty, 188
Pracchi, R., 44
Prague, 173, 174, 179
'praxiological' approach, 18
praxis, geographic thought and, 91, 92, 94
Précis de Géographie Universelle (Malte-Brun), 42
Pred, A., 76
Price, D. J. de Solla, 55
Principles of Geology (Lyell), 46

probability, 31
problem-orientation, in teaching, 34
production, Habermas's views on, 168
profit-maximization, location theory and, 181
Proudhon, P., 149
provincialism: Royce's concept of, 208, 218, 219, 222, 223; see also community
psychology, 33, 37; autobiographical approach, 97; Dilthey's influence, 101, 102, 105, 110
public exhibition, Geddes and, 194
'public geography', 30
public opinion, stages of, 77
pure geography (Reine géographie), 28
Puritanism, Royce and, 211, 212
purposive-rational action, 18
Quant, J., 210, 219, 223
quantitative methods, 32, 74, 100, 204n
'quantitative revolution', 9, 76
Radnitzky, G., 18
Rank, O., 88
Ranke, L. von, 101, 103
rationalism, 100; cf. romanticism, 167–8, 169
Ratzel, Friedrich, 62, 63, 99, 100, 110
Raumwirtschaft, 182n
Ravenstein, E. R., 159
Raynaud, Alain, 230
'reason', Hegel's notion of, 104
Rechnendes Denken, 83, 84, 90, 91
Reclus, Elie, 155, 156, 196
Reclus, Elisée, 5, 140, 196; anarchism, 156–61; circle of affinity, 14; influence on geographers, 161–3; life, 154–6
Reclus, Jacques, 154, 155, 157
Reclus, Paul, 156
Reclus, Zéline, 154

'reflection', Dilthey's definition, 103
regional development policy, 'normal' science and, 87
regional systems, logic of, 71
'regional' tradition, French, 100
regionalism, 31, 202–3
rehabilitation, Geddes's theories of, 196
Reine Théorie des Standorts, 166
'relevant geography', 73
Relph, E., 23, 100, 115, 122, 129n, 209
Repetti, Emanuelle, 44
research praxis, 17–19; and foundations of historiography, 19, 20, 21, 22
Responsible Government (Friedrichs and Cole), 183n
Révolté, Le, 144
revolution: concept of, 78; Kropotkin on, 140, 142–3
Ricardo, David, 236
Richards, I. A., 101
Richards, V., 150
Richthofen, Else von, 173, 174, 182n, 183n
Richthofen, F. von, 2, 21, 63, 100
Rickert, Heinrich, 100, 104, 112, 126n
Rickman, H. P., 127n, 128n
Ricoeur, P. 118, 123
Ringer, F., 169
Ritter, Carl, 2, 9, 29, 47, 59; and Dilthey, 110–11; founder of geographical science, 39, 40, 41, 66; as historian, 43; impact, 42; and Reclus, 155
Rivkin, M., 121
Robin, Paul, 161
Robinson, Brian S., 234, 235
Robinson, D., 211
Robinson, E. A. G., 179
Robinson, V., 150
Robson, B. T., 5
Roe, A., 77
Rogers, E. M., 77, 88, 91
romanticism, 54;

naturalistic trend in, 28; cf. rationalism, 168, 169
Roscher, W., 174, 175, 183n
Rose, Courtice, 5
Rousseau, Jean Jacques, 28
Rowles, G., 122
Royal Astronomical Society, 58
Royal Geographical Society, 42, 54, 55, 59, 60, 75; founder members, 43, 58; Kropotkin and, 143, 145
Royce, Josiah, 5, 208–10, 223; California, 220–22; elements of philosophy, 214–20; intellectual context, 210–14
Royce, S., 212
Rudner, R., 85
Ruskin, John, 191, 196
Russell, B., 33, 192
Russia, 31, 160; education, 48; institutionalization of geography in, 44–6; Kropotkin in, 134–5, 150
Russian Geographical Society, 45, 135, 144
Ryan, A., 85
Ryle, G., 110
Saarinen, T., 100
Sack, Robert D., 239n
Saey, P., 71, 182n
St Andrews, University of, 144
St Petersburg, 45, 135
Saint-Simon, Claude, 157
Sainte-Foy-la-Grande, 154
Salomon, J. J., 32
Samuels, M. S., 23, 100, 209
Sanderson, M., 43
Santayana, George, 210, 221
Santos, M., 90
Sauer, Carl, 3; Berkeley school, 71–3; Dilthey and, 100, 128n, 129n; man-land studies, 99
Scargill, D. I., 42, 59
Schaefer, F. K., 2, 39, 76, 100, 113
Schelling, Friedrich, 100, 104, 111, 169
Schiller, Johann, 101
Schleiermacher, Friedrich,

103, 108, 128n, 169
Schlüter, O., 31, 100, 110, 113–14, 128n
Schmieder, O., 110, 114
Schmittenhusen, J., 230
Schmoller, G., 173, 175
Schopenhauer, A., 173
Schroyer, T., 86
Schumacher, E. H. 138
Schutz, Alfred, 101, 108, 209
science, 37, 70, 81; continuous development of, 8; historiography, 10; history of, 22; 'internalist' conception of, 66; maritime commerce and, 46; social-cognitive nature of, 12; and society, 18, 32, 83; sociology of, 12
science policy, 32
scientific advance, 2; human implications of, 18
scientific change: nature of, 76; social dynamics of, 12
scientific content, geography and, 18
scientific knowledge, 67; and social structure, 17–19
scientific method, validity of, 227
scientific revolution: structure of, 83, 229; theory of, 81
scientific societies, institutionalization and, 29
scientific theory: emerges out of facts, 10; socio-psychological context of, 18
Scott, A. J., 165
Scott, J., 167
Seamon, David, 34, 122
self-help, Geddes's notion of, 196
Shaw, Bernard, 194
Siberia, Kropotkin in, 134, 135
Simmel, Georg, 100, 101, 110, 174
Simon, W., 169

Sklair, L., 65
slavery, 5
Smith, John, 212, 215, 216, 218
social concern, geography and, 1, 405
Social Darwinism, 137
social environment, and research praxis, 18
social history, of science, 18
'social humanism', 33
social indicators, Geddes's concept of, 197
social justice, Kropotkin's views on, 135
'social physics', 32
social planning: geographers and, 5; science and, 32
social sciences: concept of paradigm, 73; emergence of, 3; epistemology and, 228, 238; evolution of, 187; fission of, 188, 189–90, 204; political flavour, 186; social indicators and, 197
social structure: Geddes's view of, 196–8; and scientific knowledge, 17–19, 67
Société Géographique de Paris, 54, 155, 162
society: goals, 26, 30, 32; science and, 18, 32
Sociological Papers, 189
Sociological Review, 189, 205n
Sociological Society, 189, 192, 204n, 205n
sociology, 52, 190; Dilthey's influence on, 101, 110; paradigm idea, 72; racial, 189; of science, 18
Sociology, Institute of, 202
Somerville, Mary, 40, 43
Sorbonne, 41, 52
Sorokin, P. A., 86, 88
Spain, 149; education, 48
Spate, O. N. K., 100, 129n
spatial dimension, France, 165
spatial organization, Kropotkin and, 139

'spatial science', 99
spatialism, quantitative, 33
Spiegel-Rösing, I. S., 18, 32
Spottiswoode, W., 75
Stamp, L. Dudley, 76, 144, 202
Starr, L. M., 97
state, idealist conception of, 169
state support, and geographical societies, 58
statistical mechanics, and geographical reasoning, 4
statistical quantification, 31
Statistical Society, 189
statistics: and geography, 44; Russia, 45; societies, 188; see also quantitative methods
Stedman Jones, G., 168
Stirner, M., 136
Stoddart, D. R., 4, 5, 51, 59, 60, 73, 76, 81, 163, 183n, 208, 214
structuralism, in Geddes's thought, 196
Structure of scientific revolutions, The (Kuhn), 70
Struve, W., 183n
'sub-scientific line', 23
Sukhova, N. G., 45
Surveiller et punir (Foucault), 231, 234, 238
survey, Geddes and, 191, 202–3
Sweden, 50
Switzerland: education, 48, 50, 59; Kropotkin in, 136; Reclus in, 155, 156, 162
Tatham, G., 111
Taylor, Eva G., 235
Taylor, Griffith, 11
Taylor, P. J., 65, 66, 74, 76, 87, 99
Templier, Emile, 158, 159
temporality, and experience, 106
terminology, geographical, 45
Terre, La (Reclus), 155
'text': Dilthey's notion of,

117–20; 'fixed' and 'open' character of, 122–3; hierarchical structure of, 123; human geography as text interpretation, 120–24; initial credibility, 122; noetic-neomatic structure, 122
theology, 28
The Times, 144, 159
'thinking machines', Geddes and, 192–3, 203, 205n
Third Reich, Weber and, 179
Thomas, W. I., 204n
Thompson, J. Arthur, 204n, 205n
thought, reconstructing history of, 83
Thuillier, P., 10
time axis, Hagerstrand's, 19
Tolstoy, Leo, 148
topographical tradition, 21
topography, Elizabethan, 235
toponyms, localization of, 39
Törnebohm, H., 82
Toulmin, S., 13, 14, 88
Town Planning Exhibition (1910), 194
Town Planning Institute, 200
transformation juncture, man as, 19–20
Tribe, K., 73
Trivart, J., 11
Troeltsch, E., 101, 110
Tuan, Yi-Fu, 23, 100, 115, 116–17, 122, 129n, 209
Turner, Frederick Jackson, 213
Tuttle, H. N., 107, 110
Tyrwhitt, J., 192, 205n
United States, 63, 230; Davisian geomorphology, 73; Humboldt's voyage to, 40; institutionalization of geography in, 60; Kropotkin in, 144; Reclus and, 160; research traditions, 72; Royce in, 211, 213, 220; sectionalism, 218

university: appearance of geography in, 30, 59; historical-hermeneutic concerns, 169; institutionalization of new geography, 47, 54, 60, 62; institutionalized structure of science, 28, 34, 38, 65; tradition, 29
Unstead, J. F., 75
Unwin, Raymond, 200
Uppsala, 76
urban focus, Geddes and, 199–200
urban growth, 222; Germany, 175, 183n
urban locational analysis, 72
urban sociology, Chicago school, 209, 213
Urlandschaft (natural landscape), 114
utopian school, 157
Vallaux, Camille, 99
Valskaya, B. A., 45
Varenius, 2, 39
Verein für Sozialpolitik, 172, 175
Vermont, 219
Verstehen (understanding), Dilthey's concept of, 106–8, 109, 115, 116, 120, 123, 127n, 128n
Verstendun, Dilthey's concept of, 129n
Vidal de la Blache, Paul, 2, 9, 31, 63, 99, 210; epistemological depth of, 13; 'géographie humaine' tradition, 190
Vidalian school, 209
Vienna, 88
Vine, F. J., 72
Visalberghi, A., 51
Vivian, Henry, 196
vocation, 66
Volkerkunde (systematic cultural sciences), 112
voluntarism, Royce's concept of, 223
Von Engeln, O. D., 73
Wagner, H., 112
Waibel, L., 110, 113
Wann, T., 129n
Webb, Walter Prescott, 73
Weber, Alfred, 5, 67, 101, 110; cultural-intellectual crisis, 172–81; location

theory, 166ff
Weber, Helene, 173
Weber, Marianne, 172, 173, 174, 176
Weber, Max (f. of Alfred), 172
Weber, Max (b. of Alfred), 172, 173, 174–6
Wegener, Alfred, 72
Wells, A. F., 203
Wells, H. G., 187, 194
Weltanschauungen (world view), 232; Dilthey's notion of, 116
'*Weltbild*', 19, 25, 32
Weltner, G. H., 184n
Werkmeister, W. H., 214
Westermarck, E., 189
White, G., 87, 100
White, L., 210, 223
White, Morton, 210, 223
Whitehand, J. W. R., 81
Wiebe, Robert, 222
Wilhelm Dilthey: An introduction (Hodges), 125n
'will': Fichte's notion of, 104; Weber's notion of, 178
Wilson, R. J., 210, 222, 223
Wimmer, J., 113
Windelband, Wilhelm, 100, 104, 112, 210, 211, 212
Wirkungszusammenhang (study of dynamic systems), 111, 128n
Wisotzki, E., 21, 28
Wissenschaft, 18
Wittgenstein, L., 88
Wolpert, J., 100
Woodcock, G., 135, 143, 144–7, 158
World War I, 3
Wright, John K., 3, 99, 100, 128n, 224n
Wundt, Wilhelm, 101, 110, 211, 212, 220
Zeitgeist, 10–11, 83
Zeune, F., 111
Zivilisationsprozess, Weber's notion of, 177–8
zoology, 47
Zusammenhang (internal connection), Ritter's concept of, 111
Zusammenschau (unified vision), 100